Benefit-Risk Assessment in Pharmaceutical Research and Development

Chapman & Hall/CRC Biostatistics Series

Editor-in-Chief

Shein-Chung Chow, Ph.D.
Professor
Department of Biostatistics and Bioinformatics
Duke University School of Medicine
Durham, North Carolina

Series Editors

Byron Jones
Biometrical Fellow
Statistical Methodology
Integrated Information Sciences
Novartis Pharma AG
Basel, Switzerland

Jen-pei Liu
Professor
Division of Biometry
Department of Agronomy
National Taiwan University
Taipei, Taiwan

Karl E. Peace
Georgia Cancer Coalition
Distinguished Cancer Scholar
Senior Research Scientist and
Professor of Biostatistics
Jiann-Ping Hsu College of Public Health
Georgia Southern University
Statesboro, Georgia

Bruce W. Turnbull
Professor
School of Operations Research
and Industrial Engineering
Cornell University
Ithaca, New York

Chapman & Hall/CRC Biostatistics Series

Adaptive Design Methods in Clinical Trials, Second Edition
Shein-Chung Chow and Mark Chang

Adaptive Design Theory and Implementation Using SAS and R
Mark Chang

Advanced Bayesian Methods for Medical Test Accuracy
Lyle D. Broemeling

Advances in Clinical Trial Biostatistics
Nancy L. Geller

Applied Meta-Analysis with R
Ding-Geng (Din) Chen and Karl E. Peace

Basic Statistics and Pharmaceutical Statistical Applications, Second Edition
James E. De Muth

Bayesian Adaptive Methods for Clinical Trials
Scott M. Berry, Bradley P. Carlin, J. Jack Lee, and Peter Muller

Bayesian Analysis Made Simple: An Excel GUI for WinBUGS
Phil Woodward

Bayesian Methods for Measures of Agreement
Lyle D. Broemeling

Bayesian Methods in Epidemiology
Lyle D. Broemeling

Bayesian Methods in Health Economics
Gianluca Baio

Bayesian Missing Data Problems: EM, Data Augmentation and Noniterative Computation
Ming T. Tan, Guo-Liang Tian, and Kai Wang Ng

Bayesian Modeling in Bioinformatics
Dipak K. Dey, Samiran Ghosh, and Bani K. Mallick

Benefit-Risk Assessment in Pharmaceutical Research and Development
Andreas Sashegyi, James Felli, and Rebecca Noel

Biosimilars: Design and Analysis of Follow-on Biologics
Shein-Chung Chow

Biostatistics: A Computing Approach
Stewart J. Anderson

Causal Analysis in Biomedicine and Epidemiology: Based on Minimal Sufficient Causation
Mikel Aickin

Clinical Trial Data Analysis using R
Ding-Geng (Din) Chen and Karl E. Peace

Clinical Trial Methodology
Karl E. Peace and Ding-Geng (Din) Chen

Computational Methods in Biomedical Research
Ravindra Khattree and Dayanand N. Naik

Computational Pharmacokinetics
Anders Källén

Confidence Intervals for Proportions and Related Measures of Effect Size
Robert G. Newcombe

Controversial Statistical Issues in Clinical Trials
Shein-Chung Chow

Data and Safety Monitoring Committees in Clinical Trials
Jay Herson

Design and Analysis of Animal Studies in Pharmaceutical Development
Shein-Chung Chow and Jen-pei Liu

Design and Analysis of Bioavailability and Bioequivalence Studies, Third Edition
Shein-Chung Chow and Jen-pei Liu

Design and Analysis of Bridging Studies
Jen-pei Liu, Shein-Chung Chow, and Chin-Fu Hsiao

Design and Analysis of Clinical Trials with Time-to-Event Endpoints
Karl E. Peace

Design and Analysis of Non-Inferiority Trials
Mark D. Rothmann, Brian L. Wiens, and Ivan S. F. Chan

Chapman & Hall/CRC Biostatistics Series

Benefit-Risk Assessment in Pharmaceutical Research and Development

Edited by
Andreas Sashegyi

Eli Lilly and Company
Indianapolis, Indiana, USA

James Felli

Eli Lilly and Company
Indianapolis, Indiana, USA

Rebecca Noel

Eli Lilly and Company
Indianapolis, Indiana, USA

CRC Press
Taylor & Francis Group
Boca Raton London New York

CRC Press is an imprint of the
Taylor & Francis Group, an **informa** business
A CHAPMAN & HALL BOOK

CRC Press
Taylor & Francis Group
6000 Broken Sound Parkway NW, Suite 300
Boca Raton, FL 33487-2742

First issued in paperback 2020

© 2014 by Taylor & Francis Group, LLC
CRC Press is an imprint of Taylor & Francis Group, an Informa business

No claim to original U.S. Government works

Version Date: 20131009

ISBN 13: 978-0-367-57624-0 (pbk)
ISBN 13: 978-1-4398-6794-5 (hbk)

**Visit the Taylor & Francis Web site at
http://www.taylorandfrancis.com**

**and the CRC Press Web site at
http://www.crcpress.com**

Contents

Section I   Early Clinical Development

Section II   Full Clinical Development

Section III   Regulatory Review and Policy

Editors

Andreas Sashegyi   Dr. Sashegyi is currently senior research advisor in the biometrics organization at Eli Lilly and Company. He obtained a PhD in biostatistics from the University of Waterloo, Canada, in 1998 and joined Lilly later that year. Dr. Sashegyi spent the first 8 years providing statistical leadership to various compounds in late-stage clinical development, in the endocrine, cardiovascular, and critical care therapeutic areas. Dr. Sashegyi led clinical trial design and interim data monitoring strategies, and consulted on numerous regulatory interactions in support of approvals and continuing marketing authorization, involving the Food and Drug Administration as well as the European Medicines Agency. In 2006, Dr. Sashegyi joined Lilly's Decision Sciences organization, providing portfolio management and decision analytic support to various governance committees, and guiding teams in preparation for key governance meetings involving major investment decisions. The combination of his academic training and clinical development experience, coupled with his more recent leverage of decision analysis in practice, led to an increasing interest in the integrated analysis of benefits and risks in pharmaceutical research and development. Dr. Sashegyi rejoined Lilly's global biometrics organization in 2012 and currently divides his time between therapeutic compound support and work on Lilly's Benefit Risk Advisory Group.

James Felli   Dr. Felli is a research fellow at Eli Lilly and Company in the area of drug disposition. His responsibilities include developing new decision-making tools, engaging in applied research, and internal consulting. His current research interests focus on the development of new benefit-harm models and visualizations, the modeling and analysis of stochastic multi-criteria decision problems, enriching communication between analysts and nonanalysts, and developing new methods of multidimensional data visualization. Prior to joining Lilly, Dr. Felli served as associate professor of decision science with the Defense Resources Management Institute of the Naval Postgraduate School in Monterey, California. He holds a PhD in industrial engineering and management science from Northwestern University and two Master's degrees, one in management science from the Krannert Graduate School of Management at Purdue University and one in mathematics from the State University of New York at Buffalo. Dr. Felli is a fellow and current president of the Society of Decision Professionals and an active member of the Decision Analysis Society of INFORMS and the Decision Analysis Affinity Group. His research has been published in numerous professional journals, including *Medical Decision Making, Interfaces, European*

Journal of Operational Research, Military Operations Research, Health Economics, and *Decision Analysis.*

Rebecca Noel   Dr. Noel is currently a senior research scientist with Eli Lilly and Company. Since joining Global Patient Safety in 2003, she has been extensively involved in developing and promoting systematic approaches to benefit-risk assessment, both internally at Lilly and externally via the PhRMA Benefit-Risk Action Team (BRAT), where she led the pilot program to evaluate the real-world application of the BRAT Framework. She co-chaired the Next Steps Working Group (NSWG), a collaborative working group of regulators, industry scientists, and academics. When the NSWG became affiliated with the Drug Information Association (DIA), Dr. Noel assumed the chair of the DIA working group for benefit-risk assessment. She also currently serves on the Centre for Innovation in Regulatory Science's (CIRS) Task Force on Benefit-Risk Assessment and on the Innovative Medicines Initiative work package devoted to benefit-risk assessment. Dr. Noel holds a master of science in public health from the University of Kentucky and a doctor of public health in international health and epidemiology from the University of Alabama at Birmingham.

Contributors

Elizabeth Andrews   Dr. Andrews is a research epidemiologist and consultant based within RTI International. She provides consultation for pharmaceutical companies and other organizations on epidemiologic approaches to drug safety evaluation, benefit-risk evaluation, and therapeutic risk management. She also directs or advises observational research programs that evaluate drug safety and measure compliance with prescribing and treatment guidelines. These studies use existing healthcare databases or require field data collection in the United States and Europe. She helped develop the Pharmaceutical Research and Manufacturers of America-initiated Benefit-Risk Action Team (PhRMA BRAT) Framework through its implementation and refinements using a series of case studies in a partnership between RTI and the PhRMA BRAT team.

Previously, Dr. Andrews was vice president, Worldwide Epidemiology, at GlaxoSmithKline and predecessor companies, starting in 1982. Prior to joining industry, she managed the Statewide Regionalized Perinatal Care Program and directed the non-Medicaid healthcare reimbursement programs for the State Health Department of North Carolina.

Dr. Andrews is a fellow and past president of the International Society for Pharmacoepidemiology, and coeditor of *Pharmacovigilance*. She is an adjunct associate professor of the University of North Carolina School of Public Health and School of Pharmacy, and a member of the Scientific Advisory Board Observational Medical Outcomes Partnership.

Philip Bonforte   Mr. Bonforte is an assistant vice president at FaegreBD Consulting. His work focuses on FDA policy and regulatory affairs.

With a combination of legal, regulatory, and policy experience, Mr. Bonforte counsels clients on agency reforms and legislative proposals. His practice touches the complete life cycle of FDA-regulated products, from pre- to post-market.

Mr. Bonforte works closely with pharmaceutical manufacturers and patient groups on benefit-risk issues, analyzing structured assessment methodologies and advising on global trends in benefit-risk policy development. His expertise also includes supply chain integrity, FDA user fees, and drug shortages.

Mr. Bonforte holds a BA degree from Union College and a JD degree from Seton Hall University School of Law.

Don Buesching   Dr. Buesching has worked extensively in the areas of health-related quality of life and patient-reported outcomes in his 20-year career at Lilly, where he has been involved in the development of instruments in anxiety, depression, and urinary incontinence. He has served as a temporary advisor to the World Health Organization on the development of the WHOQOL, the WHO instrument to measure generic health-related quality of life. More recently, Dr. Buesching has turned to the problem of using patient-reported measures to address heterogeneity of treatment response.

One of Dr. Buesching's other areas of research interest is pragmatic clinical trial design, where he has participated in a number of pragmatic clinical trials. He currently serves as an internal consultant on pragmatic trial design in the areas of depression, schizophrenia, Alzheimer's disease, and diabetes. He has also led or participated in the execution of a number of both retrospective and prospective observational studies. Dr. Buesching received his undergraduate degree from Butler University and holds an MS and PhD in medical sociology from the University of Illinois at Champaign-Urbana.

James Cross Dr. Cross joined Genentech, Inc., a member of the Roche Group, in November 2009. Initially, he worked as a senior scientist in drug safety, where he led projects to evaluate new methods for improving benefit-risk assessment and to create and implement company guidance and training on benefit-risk assessment. These activities have included several studies on incorporating patient perceptions of benefits and risks into early-stage and late-stage product development. Since July 2011, he has worked as a program director in regulatory affairs, where he has led U.S. and global regulatory strategies for various oncology projects. In 2009, Dr. Cross completed his PhD thesis, titled "Novel Quantitative Methods for Drug Benefit-Risk Assessment in Regulatory Decision-Making: A Case Study and Policy Analysis." From 2000 to 2004, Dr. Cross worked at the FDA, first in the Division of Metabolic and Endocrine Products, and subsequently in the Office of New Drugs. Prior to the FDA, he completed a fellowship at the Center for Drug Development Science under former FDA center director Carl Peck to evaluate patterns in labeled changes to drug dosages. He completed his bachelor's degrees in biology and French from Tufts University, Master's degree in immunology from the University of California at Davis, and PhD in pharmaceutical outcomes research from the University of Washington. He has published and presented internationally on the topics of benefit-risk assessment, drug safety, and regulatory policy. Dr. Cross is an active member of the Drug Information Association's Special Interest Area in Benefit-Risk Assessment, and the Biotechnology Industry Organization's Pharmacovigilance and Benefit/Risk Management Working Group.

John Ferguson Dr. Ferguson is vice president and global head of medical safety and pharmacovigilance at Novartis Vaccines. A board-certified cardiologist and former co-principal NIH and MRCC investigator, he received his training in cardiology and clinical epidemiology at McGill University, McMaster University, and Cedars-Sinai Medical Center. Prior to entering the pharmaceutical industry, his research focused on expert systems, risk prediction in coronary artery disease, and development of catheter-based interventions for the prevention and treatment of ischemic stroke. During more than 15 years as a global head of clinical safety and risk management, he has contributed to the approval and benefit-risk optimization of small molecules, biologics, and vaccines. He is an active participant in industry organizations, having served on clinical trial, safety, risk management, and benefit-risk management committees and working groups at PhRMA and BIO.

Dr. Ferguson is a faculty member in the Harvard Medical School Collaborative and Distance Learning Program in Clinical Research, a long-standing contributor to the Massachusetts Institute of Technology's Center for Biomedical Innovation (MIT-CBI), and has served as a panelist and key subject matter expert for the Institute of Medicine (IOM), as well as on working groups for the Council for International Organizations of Medical Sciences (CIOMS) and the Bill and Melinda Gates Foundation.

His current work focuses on adaptive licensing and developing scientific and structured approaches to characterizing, optimizing, and communicating the benefit-risk balance of medicinal products.

Timothy Franson Dr. Timothy Franson is a principal in FaegreBD Consulting's health and biosciences sector and leads the firm's regulatory affairs practice. This group provides traditional comprehensive preapproval and post-marketing regulatory advice. He has extensive clinical and regulatory experience in all pre- and post-approval phases of pharmaceutical development (small and large molecule) relating to interactions with the FDA for policy and product issues, as well as interactions with global regulators.

Dr. Franson was the founder and president of Franson PharmAdvisors LLC, a clinical and regulatory pharmaceutical development consulting firm. The company merged with FaegreBD Consulting when Dr. Franson moved from senior advisor to senior vice president at FaegreBD Consulting. Before joining FaegreBD Consulting, Dr. Franson was vice president of Global Regulatory Affairs at Lilly Research Laboratories (Eli Lilly and Company), responsible for all regulatory and patient safety activities from 2003 until his retirement from Lilly in June 2008. He joined Eli Lilly and Company in 1986. From 1997 to 2003, Dr. Franson was vice president of Clinical Research and Regulatory Affairs-U.S. for Lilly. In regulatory affairs from 1995 to 2008, Dr. Franson was directly responsible for Lilly's FDA submissions (NDAs, supplements), which involved more than 20 major submission reviews and approvals, as well as regulatory compliance and policy matters.

Dr. Franson has been a leader in many industry initiatives. He was co-chair of the joint FDA-industry working group addressing clinical aspects of the FDA Modernization Act of 1997, including the Prescription Drug User Fee Act (PDUFA) renewal. From 2000 to 2003, he co-chaired industry-FDA committees for PDUFA-3 renewal and has testified, on behalf of industry, at several congressional hearings. Dr. Franson also co-chaired an FDA-industry safety interventions working group, was a member of the AAMC-PhRMA Clinical Trials Forum, and was a member of the Regulatory Advisory Board for the Centre for Medicines Research International from 2003 to 2008. Franson recently served on the NIH-NCATS Treatment of Rare Diseases (TRND) review panel. From 2009 to 2013, he served on the board of directors at Myrexis Pharmaceuticals (MYRX, NASDAQ). Franson is currently the president of the U.S. Pharmacopeial Convention (2010–2015).

Dr. Franson has authored more than 50 articles and has lectured extensively in the fields of infectious disease, epidemiology, pharmacoeconomics, benefit-risk management, and antibiotic utilization.

Alicia Gilsenan   Dr. Gilsenan is a research epidemiologist within RTI International, and is also a licensed pharmacist. Her primary area of expertise is pharmacoepidemiology and therapeutic risk management. Since joining RTI in 1997, she has directed multidisciplinary international teams for large primary data collection and database studies in the United States and Europe. Currently she leads an international safety surveillance study and a U.S.-based patient registry that are both part of a mandated risk management program and is involved in several Risk Evaluation and Mitigation Strategy (REMS) projects. She helped develop the Pharmaceutical Research and Manufacturers of America-initiated Benefit-Risk Action Team (PhRMA BRAT) Framework through its implementation and refinements using a series of case studies in a partnership between RTI and the PhRMA BRAT team. Dr. Gilsenan has published within multiple therapeutic areas, including mental health, oncology, sexual functioning, cardiovascular health, respiratory health, and smoking cessation, as well as risk management and benefit-risk assessment.

Dr. Gilsenan is adjunct assistant professor in the School of Pharmacy at the University of North Carolina at Chapel Hill.

A. Brett Hauber   Dr. Hauber is a senior economist and the vice president of Health Preference Assessment at RTI Health Solutions. He has more than 15 years of academic, research, and government experience in health and environmental economics. His primary area of specialization is in conducting conjoint analyses and discrete choice experiments to quantify preferences for medical interventions and health outcomes. He also has extensive experience in conducting benefit-risk analysis of patients and other healthcare decision makers, estimating willingness to pay and examining the relationship between patient preferences and adherence. He has studied the theoretical and empirical relationships among various health utility measures. His most recent applied work has included discrete choice experiments of patient and physician benefit-risk preferences for treatments for conditions in numerous therapeutic areas, including neurology, infectious diseases, women's health, gastrointestinal diseases, diabetes, and oncology. He has also studied the theoretical and empirical relationships among various health utility measures. Dr. Hauber regularly teaches courses on conjoint analysis. He currently is the chair of the International Society for Pharmacoeconomics and Outcomes

Research (ISPOR), Conjoint Analysis–Statistical Analysis, Reporting, and Conclusions (CA-SARC) Task Force, and was previously a member of the ISPOR task force that developed the ISPOR Checklist for Good Research Practices in Conjoint Analysis. Dr. Hauber's research has been published in numerous health outcomes and medical journals.

F. Reed Johnson   Dr. Johnson has more than 35 years of academic and research experience in health and environmental economics. He has served on the faculty of several universities in the United States, Canada, and Sweden. He currently holds the position of distinguished fellow and principal economist at RTI Health Solutions. As a staff member in the U.S. Environmental Protection Agency's environmental economics research program during the 1980s, Dr. Johnson helped pioneer development of basic nonmarket valuation techniques. These techniques are now widely used for cost-benefit analysis in health and environmental economics. He has designed and analyzed numerous surveys for measuring willingness to pay for health risk reduction and improved environmental quality. He also has developed and researched various value elicitation methods, including direct questions, discrete choice, graded pairs, and other approaches. Dr. Johnson has extensive experience in using advanced statistical techniques to analyze censored, truncated, and discrete survey data.

Dr. Johnson has more than 100 publications in books and peer-reviewed journals. His research has been published in various clinical journals, including *Review of Economics and Statistics, Journal of Health Economics, Medical Decision Making, Health Economics, Value in Health, Journal of Environmental Economics and Management, Journal of Policy Analysis and Management,* and *Land Economics.* He has coauthored a book on techniques for using existing environmental and health value estimates for policy analysis. His current research involves estimating general time equivalences among health states and patients' willingness to accept side effect risks in return for therapeutic benefits.

Joseph Johnston   Dr. Johnston is a research fellow at Eli Lilly and Company in the Global Health Outcomes research function. Over the past decade, he has held a variety of positions, including scientific lead for phase IV outcomes research for multiple compounds, leader of an epidemiology team responsible for research supporting phase II/III product teams across the portfolio, and leader of a health technology assessment team responsible for developing core reimbursement dossiers and economic models for use by Lilly affiliates globally. His current research interests focus on applications of clinical epidemiology, decision analysis, pharmacoeconomics, and evidence-based medicine to better understand the clinical effectiveness and value of pharmaceuticals. Prior to joining Lilly, he completed a residency in internal medicine and pediatrics and a fellowship in health services research at the Cincinnati Children's Hospital Medical Center and the University of Cincinnati Hospital, where he serves a volunteer assistant professor of clinical medicine. Dr. Johnston holds an MS degree in epidemiology from the Harvard School of Public Health, an MD degree from the University of Missouri at Columbia, and an MS degree in computer science from the University of Illinois at Champaign-Urbana. He is a fellow of the American Academy of Pediatrics, and a member of the American College of Physicians and the Society for Medical Decision Making. His research has been published in numerous professional journals, including *Medical Decision Making, Archives of Internal Medicine, American Journal of Medicine, Journal of General Internal Medicine, Medical Care*, and *Quality of Life Research.*

Bennett Levitan   Dr. Levitan is director of Quantitative Safety Research in the Department of Epidemiology at Janssen Pharmaceutical Research and Development. He is also an adjunct professor at the Center for Bioinformatics and Computational Biology at the University of Delaware. He has over 20 years of experience in decision analysis, modeling, and simulation in both consulting and pharmaceutical arenas. He specializes in pharmaceutical benefit-risk assessment and is a frequent speaker at meetings on the topic. Dr. Levitan designs and oversees the development of benefit-risk analyses for use in advisory committee and health authority submissions worldwide, for compounds both in development and post-approval. His experience spans many therapeutic areas, with an extensive focus on cardiovascular health and schizophrenia.

Dr. Levitan co-led technical development of the PhRMA Benefit-Risk Action Team (BRAT) Framework for drug benefit-risk assessment, currently being used by several pharmaceutical companies. He is a member of the Next Steps Working Group, a public-private partnership focused on methods and collaboration for pharmaceutical benefit-risk assessment, and is also a member of the Centre for Innovation in Regulatory Science (CIRS) Benefit-Risk Task Force and the International Society for Pharmacoepidemiology (ISPE) Benefit Risk Assessment Special Interest Group.

Dr. Levitan received his BSc (electrical engineering) from Columbia University in New York and his MD-PhD (bioengineering) from the University of Pennsylvania. He worked as a postdoctoral fellow at the Santa Fe Institute. His research and consulting work have dealt with pharmaceutical benefit-risk assessment, organizational learning, evolutionary-based optimization, high-dimensional data visualization, and combinatorial chemistry.

Marilyn Metcalf   Dr. Metcalf is the senior director of Benefit Risk Evaluation at GlaxoSmithKline. She leads GSK's work to understand the balance between the favorable and unfavorable effects of their medicines. The BRE team and their colleagues use benefit-risk analysis and visualization to provide insights that improve treatments for patients.

Previously, Dr. Metcalf worked in reproductive health research while completing her doctorate in sociology at University of North Carolina–Chapel

Hill. As a project director at Family Health International, she designed and analyzed contraceptive clinical trials and HIV vaccine research primarily for developing countries. She moved to the former GlaxoWellcome to study the health economics and quality of life effects of HIV therapies. After the GSK merger, she moved to the UK to build GSK's international Decision Sciences team. Upon returning to the United States, she led Centocor's R&D portfolio management team and directed GSK's Quantitative Decision Sciences department. She began her current role in 2011. Marilyn is an active member of the Public-Private Benefit Risk Working Group (formerly NSWG), the Centre for Innovation in Regulatory Science's benefit-risk technical advisory team, the European Innovative Medicines Initiative's work on benefit-risk methodology, the Society of Decision Professionals, and PhRMA's benefit-risk working groups.

Lawrence D. Phillips   Dr. Phillips is an emeritus professor of Decision Sciences at the London School of Economics (LSE) and a director of Facilitations Limited. When he is not teaching in the LSE's MSc Program in Decision Sciences, he works as a process consultant, helping key players in organizations facing decisions that balance risk, uncertainty, and multiple conflicting objectives. From 2009 to 2011 he led the European Medicine Agency's Benefit-Risk Project, which explored how explicit, quantitative methods and tools could provide assistance to decision makers, and he is similarly engaged with the Innovative Medicines Agency's PROTECT project, which ends in 2014. He has worked with 17 pharmaceutical and biotechnology companies over the past 27 years, applying decision and risk analysis to issues of strategic and operational management, option appraisal, benefit-risk assessment, project prioritization, portfolio management, resource allocation, and crisis management, covering all stages in the lifetime of drugs, from discovery and research through development, marketing, and post-marketing.

Dr. Phillips has authored over 120 publications on behavioral decision theory, decision and risk analysis, Bayesian statistics, organization theory, group processes, and cultural differences in dealing with uncertainty. He has served on the editorial boards of *Acta Psychologica*, the *Journal of Behavioral Decision Making*, the journal *Decision Analysis*, and continues to serve on the *Journal of Forecasting*. In November 2005, the Decision Analysis Society awarded Dr. Phillips the Frank P. Ramsey medal for "distinguished contributions to decision analysis."

Ralph Swindle   Dr. Swindle, a Lilly research fellow, has been with Lilly Global Health Outcomes research since 1999. He came to Lilly after 12 years in VA Health Services Research and Development (HSR&D) at the Palo Alto VA/Stanford University Medical Center and the Indianapolis (Roudebush) VA/Indiana University Medical Center. He is currently the technical lead across all products for U.S. Health Outcomes and Health Technology Assessment.

He is part of the Global Health Outcomes Center of Excellence as a lead reviewer of observational studies for EXPERT (the Excellence in Pragmatic and Observational Studies Research Design Network), the scientific advisor to the U.S. payer customer rapid response research process, and the lead for U.S. outcomes-based contracting research. Dr. Swindle has an extensive background in outcomes research, health services research, and psychometrics (Patient Reported Outcomes development). He is widely published in health services, medical, and psychological journals, with over 80 peer-reviewed journal publications and 10 book chapters; he has made over 100 conference presentations. His most recent publications have been in the areas of chronic low back pain, depression, diabetic neuropathic pain, mental health screening in primary care, hypogonadism, fibromyalgia, and insomnia.

Jing Zhang   Dr. Zhang is a business reporting analyst at 1800PACKRAT and an adjunct professor at the Department of Bioresource Policy and Business and Economics at the University of Saskatchewan, Canada. She was a research economist at RTI Health Solutions when she coauthored the chapter on quantifying patient preferences, contained herein. Dr. Zhang has nearly 10 years of research experience using stated preference methods to elicit individuals' preferences for nonmarket goods and services such as health treatments, novel food products, and environmental goods and services. She has developed expertise in choice experiment design and discrete choice modeling. Dr. Zhang has published papers in *Land Economics, Journal of Environmental Economics and Management, Journal of Agricultural Economics,* and *Empirical Economics.*

Prologue

This book is about benefit and risk.

More precisely, this book is about the definition, assessment, balance, and communication of favorable and unfavorable consequences of treatments in pharmaceutical research and development.

This book is not a definitive treatise on benefit and risk in pharmaceutical research and development. It is not a prescriptive guide for characterizing the benefits and risks of a molecule under development. Nor is it a collection of best practices and recommendations for a successful benefit and risk assessment.

It is a gateway.

It is a gateway that opens into a long corridor that chronicles the concepts, assessment methods, interpretations, and implications of benefits and risks as a molecule journeys from concept to customer. Along the corridor are four doors that open into four galleries. Depending upon our experiences and state of mind, these doors may appear dark and foreboding, etched with esoteric runes and rubbed with mystic herbs, or curious and unusual entryways we've walked by for years yet never thought to enter, or enticing and accessible points of access to newly revealed vistas abundant with challenge and promise.

Each door in the corridor will bear one of four brass plates: Early Clinical Development, Full Clinical Development, Regulatory Review and Policy, or Post-Launch Assessment. In the Early Clinical Development gallery, we will encounter interpretations of benefit and risk in the context of a molecule moving from discovery through its preclinical evaluation and its initial testing in man. The Full Clinical Development door opens into a gallery that considers benefit and risk during a molecule's journey from its entry into man until it is submitted to regulators for approval. The Regulatory Review and Policy gallery offers insight into the role of benefits and risks in guiding decisions that promote predictable, rational outcomes, as defined, implemented, and enforced by authoritative agencies. The post-launch door opens into the final gallery, in which the molecule becomes a drug available for consumption by people, and efficacy fades in the wake of effectiveness.

Within each gallery, one or more beautifully rendered paintings will be proudly presented. Although they vary in style and brushstroke, perspective and color, each painting portrays an important vision of drug development germane to its gallery. Each painting's artists will stand to meet us beside their work, to interpret their vision as words on a page, and to anticipate our questions and offer answers. The questions posed and answers provided will reveal the artists' interpretation of benefit and risk, their assessment and characterization of benefit-risk, and their assertions about its relevance and interpretation in the context of their understanding, expertise, and experience.

Just as our perception of each door was shaped by our experiences and state of mind, so too are the lenses through which we will view the paintings in each gallery. Our lenses may be jaded, dark, and smeared with the grime of bias and resolute preconception, or scratched and chipped by a series of misperceptions or misconceptions, or crystal clear, but new and slightly blurry, requiring some time to get used to the new focal points they exercise. Whatever the state of our lenses, together we will remove them and look at the paintings with fresh eyes and untouched vision.

Each painting in each gallery was carefully commissioned by artists well recognized as experts in their field, actively engaged in the development of benefit-risk methods and applications, and erudite in the areas of benefit-risk assessment, interpretation, and communication. They are scientists and professionals, academics and practitioners, whose collective expertise touches upon all aspects of benefit-risk from concept to modeling to interpretation to policy. Their art is presented with pride, for the purpose of affecting both a cognitive understanding and an emotional appreciation of the critical importance of benefit and risk in the design, development, and delivery of pharmaceuticals to people in need. Their biographies are available in the front matter, should one desire to commission a dedicated piece for their home or office.

And so we stand before the gateway and prepare to walk through.

Before we take that first step, however, we must clarify three things. First is the issue of benefit and risk. In essence, we use these terms to capture the collections of desirable and undesirable consequences of treatment: benefits are favorable effects we wish to secure, and risks are unfavorable effects we wish to avoid. We must recognize, however, that the term *risk* is not restricted to this meaning, and carries with it both the chance and consequences of ill fortune. We face the risk of death from a car accident every time we venture out in our cars; we face the risk of manifesting disease from the genetic mix we carry in our cells; we face the risk of rejection each time we open ourselves to another human being. And when it comes to pharmaceuticals, we run not only the risk of suffering adverse events, but also the risk that the treatment will not work as expected for us, and that the benefits we receive might fall short of our expectations. The concept of risk—the chance of ill fortune—lies on both sides of the benefit-risk statement.

Second is the issue of language. We have tried to use a common language in this book, and have worked with our authors to present words such as *benefit* and *risk* in common terms that transfer from chapter to chapter. Nevertheless, we are constrained by terminology developed and calcified over decades of use. Within this book, we will use the terms *benefit* and *risk* as they have come to be employed in the pharmaceutical industry. On the one hand, benefits are favorable outcomes and desired consequences expected from treatment; on the other hand, risks are unfavorable outcomes and undesired consequences that might arise from treatment.

And so, as we stand before the gateway, we hold our two hands out: on our right hand, a set of benefits we wish to secure; on our left, consequences to be avoided, mitigated, and managed—at all times cognizant of which hand holds the more weighty collection. Each consequence we lay upon a hand becomes weightier, for good or for ill, as its likelihood and magnitude of effect increases, and our hands become the pans of a scale balance. And so we come to the notion of a benefit-risk balance and the implication that we seek to bring our left and right hands into balance. But this is certainly not the case: given our druthers, every scientist, manager, marketer, analyst, technician, physician, caregiver, payer, regulator, and patient would prefer to have their right hand crushed under the enormity of desired outcomes and their left as light as a feather, devoid of even the most benign of potential inconveniences. Yet such is rarely, if ever, the case. With the potential for benefit comes the potential for harm, for the pharmaceutical agent is active in the human body and can trigger effects both predictable and not, theoretical and observed. So the left hand too is weighed down, and the question becomes: According to our scale balance, how much weight on the left hand can be acceptably offset by weight on the right?

And so comes our third point of clarification: perspective. Who gets to define "acceptable"? From the perspective of a regulator, even very low probability events will obtain given a sufficiently large population. How many people do we condemn to suffer by our approval? From the perspective of a payer, some undesirable consequences may lead to very costly future states for large groups of patients. How many people must benefit to justify the costs consequent to the addition of a treatment to a package of goods offered? From the perspective of a physician, averages lose their meaning and modes become important as focus shifts to specific patients under the constraint of doing no harm. How do we communicate that information? From the point of view of a patient, it all comes down to personal risk tolerance and willingness to gamble. How can I understand what it takes to "beat the odds?"

So, here we stand. Hands out. Arms open to recognize not only the balance we seek to understand, but to signify our openness to consider new ideas and entertain new visions as we peruse the galleries before us. We can take our time: the galleries are open solely for our benefit, and each beckons with its own allure. In time, we will visit them all, some more than once, knowing that the whole experience is necessary to understand the story that each may tell only in part.

That is what this book is about.

This is a book about an evolving area in pharmaceutical research and development: the concepts, assessment methods, interpretations, and implications of favorable and unfavorable consequences of treatments as a molecule journeys from concept to customer.

This book is a journey of discovery.
This book is about benefit and risk.

Note on Arrangement of Content

The material in this book has been arranged so as to follow the chronological order of drug development, beginning from early clinical development, through late-stage development, regulatory review, and finally post-launch assessment. Whereas this ordering is logical and provides an intuitive flow for reading the book from front to back, it does not imply that the sections cannot just as easily be taken out of order. An alternate, equally viable arrangement of this book would place the "Regulatory Review and Policy" section first. This section discusses the policy considerations around benefit-risk analysis, and provides a good overview of the evolution up to the present of benefit-risk analysis in the context of regulatory agencies. Thus, this section could be argued to be foundational to the rest of the book. In any case, each section is written to stand on its own, and the book was compiled in recognition of the fact that different readers will focus their attention on different parts.

Disclaimer on Content

The views expressed in this book are the views of the contributing authors and not necessarily those of the companies or agencies they represent.

Section I

Early Clinical Development

Introduction

The handle of the door to the Early Clinical Development gallery is shiny and new, and the gallery itself is spacious and open. It is also empty of artwork. The walls are bare, freshly painted, and awaiting adornment. Except one. On the far wall are two metal hooks awaiting the hanging wire of a frame. Beneath the hooks, leaning against the wall, is an empty, gilded frame. A few feet in front of the frame is an unfinished painting on a large, wooden easel.

Different artistic styles and application techniques suggest the painting to be the work of many hands, yet the emergent image is clear: a winding path through a convoluted, subtly oppressive landscape. The basic elements are in place—gnarled trees and tangled brush, nettles and vines, sprays of light through a dark canopy—but lack detail. Several large areas of canvas still await paint.

Next to the easel is a table laden with palettes fresh and used, tubes of paint, cans of turpentine and alcohol and linseed oil, and squat pots bristling with brushes. The table has been dirtied by use and cleaned afterward many times. The last user appeared to favor blue.

There is an opened envelope on the table, and a card partially removed from it. The card has been removed from and returned to the envelope many times. It is an invitation to paint.

1

Pharmaceutical Benefit-Risk Assessment in Early Development

Bennett Levitan and James Cross

CONTENTS

1.1 Introduction: Why Undertake Benefit-Risk Assessment in Early Development?

The initial thoughts one might have on benefit-risk (B-R) assessment are that it requires a detailed understanding of the various benefits of a treatment, extensive knowledge of the associated safety issues, and large quantities of detailed clinical trial or observational data to support the corresponding measures and assessments. In contrast, the early phases of clinical development, typically phase 1 and the earlier or proof-of-concept portions of phase 2, are generally characterized by lack of clarity of the relevant endpoints to measure, uncertainty about the sizes of the expected effects or even whether there will be effects at all, and little information about safety issues. Even for drugs with well-established mechanisms of action, where there is good understanding of the types of benefits and harms expected, the limited data available are subject to considerable uncertainty. The gulf between the information available in early development and the detailed information needed for B-R assessment from the perspective described above seems insurmountable, and the regulatory decision is often years ahead. Is there any motivation for companies to invest time and resources into structured approaches for B-R assessment during early development? What is the value proposition for early development B-R assessment?

Structured approaches to B-R assessment involve considerably more than quantifying, weighting, and displaying endpoint data. The qualitative and semiquantitative steps for B-R are stressed in projects and initiatives by regulatory groups, industry groups, and public-private partnerships, including the U.S. Food and Drug Administration (FDA), European Medicines Agency (EMA), International Conference on Harmonization (ICH), Pharmaceutical Research and Manufacturers of America (PhRMA), and Center for Innovation in Regulatory Science (CIRS).[1-10] Steps such as establishing an assessment or decision context, getting stakeholder agreement on the endpoints for the assessment, and investigating the degree to which the planned trials will support those endpoints all are critical parts of constructing a B-R assessment and can be of considerable value in early development. Our view is that B-R in early development is a preliminary version that helps direct considerations for the design of phase 3 trials and downstream B-R assessment, but that the same principles apply throughout the development process. This preliminary approach to B-R can also serve as a basis for end-of-phase 2 discussions with health authorities in an effort to achieve agreement on how B-R will be assessed during phase 3 studies and in marketing authorizations or new drug applications.

Other applications of B-R assessment in early development stem from its potential to serve as a structural link between a target product profile (TPP) and the clinical development plan, its utility in contributing to internal "stage gate" decisions, and its use in suggesting minimally important effect sizes or rates needed for benefits to outweigh harms. As health authorities continue to advance proposals for structured approaches toward benefit-risk,[1,5,10] sponsors will increasingly need to get an earlier start on preparing their study designs and analyses to support these growing requirements.

1.2 Differences between Benefit-Risk Assessment in Early and Full Development

One of the primary distinctions between B-R assessments performed in early versus full development is the limited availability of clinical data and the attendant degree of uncertainty. Some of the most prominent uncertainties in early development include lack of clarity about:

- *Which endpoints are important to include in the assessment.* Especially in the earlier parts of early development, clinical evidence is often extrapolated from preclinical data in animal studies or from studies in healthy humans. The benefits or harms seen in animals may not appear in humans, may appear in humans without prior preclinical evidence, or may present differently. It may also not be clear whether the benefits or harms are dose dependent. Rare safety events are very unlikely to be known or even suspected in early development.

- *The most informative way to assess or measure an endpoint.* There are often several types of measures that can be used for the same objective. For example, when considering an efficacy endpoint of a treatment for a viral infection, time to virus clearance, percent of patients with virus clearance by a given time, and percent of patients without a set of symptoms are all reasonable possibilities. However, each may dictate a different B-R assessment approach. Biomarkers may also prove misleading as endpoints. For example, decreased cellular metabolism is a biomarker for the activity of anticancer agents, but it may not correlate with significant improvements in progression-free or overall survival measured in mid-stage or late-stage trials.

- *The sizes of treatment-related effects.* The effect sizes in early development are usually based on small samples and are subject to considerable uncertainty. Additionally, potential changes in the study design between phase 2 and phase 3, such as in the inclusion/exclusion criteria and dose used, may alter the relevance of the data that are known.

- *The definition of a "responder" versus a "nonresponder,"* a challenge of considerable relevance to targeted drug discovery and personalized medicine.

Many of these challenges also occur in full development, as uncertainty never fully resolves in drug development, but their impact on decision making and trial results is considerably greater in early development.

Another major difference between B-R assessment in early and full development relates to the speed of development. Full development studies typically require several months to several years to complete, especially with long-term follow-up, while early development studies are typically done in a few weeks to a few months. The short timelines in early development limit the types of B-R analyses that can be performed. Approaches that use ancillary studies that require months to prepare and conduct, such as patient survey-based studies, may not be realistic in the given time frame, though they may have been used in developing target product profiles and suggesting target endpoints. Such studies may be premature for early development B-R assessment, since they typically are based on a short list of benefits and harms that can easily become partially or fully irrelevant once the next round of clinical studies is complete.

1.3 Applications of Benefit-Risk Assessment in Early Development

Despite the challenges, we submit that there are a number of important benefits from using a structured approach to B-R assessment in early development. The key benefits reside in the qualitative aspects of a B-R framework: getting consensus on a context for the assessment or subsequent decision, identifying the outcomes for the assessment, developing draft means to communicate the assessment, and connecting the B-R assessment with the target product profile. While some of these steps are not strictly necessary, their uptake will facilitate integration of B-R assessments into internal stage gate decisions (e.g., end-of-phase 2), assist with planning for later phase studies, and help teams achieve agreement with health authorities on an approach for B-R assessment for regulatory review.

1.3.1 Assessment or Decision Context

Even with well-integrated groups working on the same drug or device, team members often disagree on what specific decisions they face and the overall context of the benefit-risk assessment. In general, the context is the set of administrative, political, and social issues related to the decisions. It includes the goals and perspectives of the body performing the assessment or making specific decisions, historical context, and the identification of those affected by its consequences.[11] Specific objectives or questions to be considered, such as a critical B-R trade-off or important subgroups to consider, should be stated explicitly as part of this initial step.

In B-R assessments for drug development, the assessment or decision context has a relatively consistent structure and typically includes:[6,7,10]

- Nature of disease
- Medical need
- Disease and treatment epidemiology
- Study treatment, dose, and formulation
- Indication(s)
- Patient population with key inclusion/exclusion criteria and critical subgroups

- Comparator(s)
- Time horizon for outcomes
- Perspective of the involved parties, including decision-making bodies

Despite the apparent basic nature of this list of items, getting team agreement on them early in the development process is of surprisingly great value. Teams often debate the comparator, the time horizon, the exact population, and other items at length—even after considerable time and resources have been invested in a development program. Since there are often multiple stakeholder groups affected by a B-R assessment, this exercise also has value in helping highlight differences in understanding between the team performing the assessment and other stakeholders (physicians, patients, regulators, advisory committee panel members, payers, etc.). Without a clear context, other steps in B-R assessment and development in general, including outcome identification and data collection, may not be well focused and could go astray.

1.3.2 Identifying Attributes

Perhaps the most critical activity in B-R assessment, not just in early development but throughout a product's life cycle, is the specification of the full set of attributes to use in the assessment. These attributes are generally clinical endpoints or patient-reported outcomes, but may also include other properties, such as formulation, adherence, and the need for regular monitoring, although these attributes may have a lesser role in regulatory review. In some circumstances, such as the development of drugs for diseases for which there is no currently approved (authorized) therapy, additional methods, such as conjoint analysis, have been explored to identify attributes of benefit or risk by reaching out to caregivers or other stakeholders.[12,13]

In structured B-R methodologies, identifying attributes typically starts with a facilitated meeting of clinical, statistical, medical affairs, and other relevant functions (e.g., outcomes research, commercial development planning, marketing). The discussion starts with a large pool of candidate outcomes based on historical and recent clinical studies, observational studies of the proposed treatment population, target product profiles, meta-analyses for related treatments, regulatory reviews, health technology assessments, and other sources. Formal facilitation is not required, though it often proves helpful.[11,14] The team considers which attributes are relevant, measurable, important, and can collectively characterize the treatments sufficiently for clinical and regulatory decisions. The attributes are not simply labeled, but are defined with the same rigor as typically used in a statistical analysis plan. The result is a well-defined set of attributes that will be used in the B-R assessment, along with documented rationales for why different attributes were included or excluded. This step can be considered an adjunct to the process whereby product development teams identify primary and secondary outcomes for clinical trials (though B-R considerations may require modifications to the endpoints traditionally used in certain therapeutic areas).

More succinctly, the goal is to select and define prospectively the benefit and risk outcomes to be considered by (1) identifying the pool of candidate outcomes for the assessment, (2) deciding which outcomes to include in or exclude from the assessment, and (3) documenting all critical assumptions for these inclusion and exclusion decisions.[6,7] A tool that is often valuable for identifying outcomes for B-R assessment is an attribute tree. In the decision analysis literature, attribute trees are generally used to represent the goals, objectives, or values that decision makers have for a decision problem. The actual measurements for these objectives are called attributes. In pharmaceutical B-R applications, the trees lead directly to the measurements, and the similarly used terms *objective hierarchy* and *value tree* can lead to confusion, since the goals of the decision are not directly represented in the tree. For these reasons, the trees are referred to as attribute trees in this book. An attribute tree is a visual, hierarchic display of the key ideas, attributes, or criteria relevant to the decision.[11,14,15] Figure 1.1 shows an exemplar attribute tree for a cardiovascular treatment.

Building an attribute tree during discussions can be useful in helping team members and decision makers clarify which benefits and risks are pivotal to the B-R balance. The visual nature of an attribute tree greatly enhances communication—potentially serving as a means to facilitate agreement between a sponsor and a health authority for a B-R approach. Our experience within

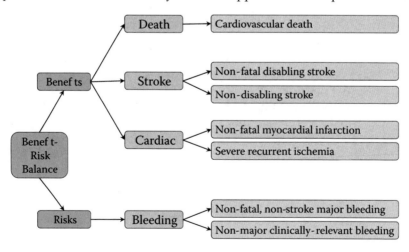

FIGURE 1.1
Example value tree for treatment of acute coronary syndrome (ACS). ACS results from obstruction of coronary arteries and presents as myocardial infarction (MI) or unstable angina. Treatment, generally with anticoagulants, is focused on preventing subsequent strokes, MIs, or death. The risks are various degrees of bleeding. In this tree, attributes are defined so there is no double counting; i.e., an event will fall under one attribute only.

the team that developed the PhRMA Benefit-Risk Action Team (BRAT) Framework also suggests that the attribute tree reduces the time needed for people new to an ongoing assessment to get oriented and start contributing to discussions.

An attribute tree exercise can also help mitigate issues that arise when interpreting trial results later. For example, in Figure 1.1, the attributes (only clinical endpoints in this case) are defined to avoid double counting an event between different attributes. This makes interpretation of "events caused" and "events prevented" very clear. If the attributes instead included "cardiovascular death" under efficacy and "major bleeding" under harms, which is typical of many cardiovascular studies for antithrombotic agents, fatal bleeding events and hemorrhagic stroke events may be counted under both. By displaying all attributes simultaneously, the attribute tree enables us to recognize and decide how to accommodate such issues.

There is no single correct attribute tree for any given B-R application. Different attribute tree structures allow teams to visualize and communicate B-R assessments in different ways. The tree in Figure 1.1 has both ischemic and hemorrhagic strokes under the same two endpoints—nonfatal disabling stroke and nondisabling stroke. This provides the advantage of grouping strokes in a fashion that relates to the clinical experience of patients—nonfatal disabling strokes under one endpoint, nondisabling stokes under another, and fatal strokes grouped with all fatal events under a third. However, an equally viable alternative value tree might separate attributes for harmful events prevented by the treatment (ischemic events) and those caused by the treatment (hemorrhagic events). The tree in Figure 1.2 shows this separation

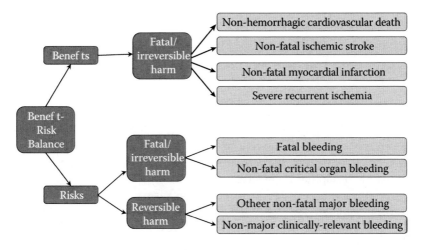

FIGURE 1.2
Alternate exemplar attribute tree for treatment of acute coronary syndrome (ACS). All ischemic events are classified as benefits and all hemorrhagic events are classified as harms, providing a separation that aligns with the treatment's mechanism of action. Additionally, benefits and harms are classified by whether they are fatal or generally result in irreversible harm.

and additionally classifies endpoints by whether they are fatal or generally cause irreversible harm. While people may disagree on the irreversible harm classifications of specific events, the additional advantage of this type of attribute tree is that it enables the comparison of groups of endpoints by their clinical impact.

Both attribute trees in Figures 1.1 and 1.2 present viable and reasonable means to set up a B-R assessment. Either or both can be used with the same audience, depending on the points being made and the perspective needed.

1.3.3 Communicating the B-R Assessment

1.3.3.1 Striving for Common Terminology

B-R assessment entails the intersection of numerous types of information: epidemiology, medical need, efficacy, safety, risk mitigation, regulatory policy, government policy, health economics, and statistics. Before considering the actual display and communication of the assessment, we have found it helpful to strive for a common language with stakeholders in these areas—to know that they share the same interpretation of the terminology used in the assessment. For example, does *comparator* indicate a particular treatment at a particular dose or a class of drugs? Are the strokes described in Figure 1.1 all strokes or only ischemic strokes? Are clinical attributes characterized in person-years or Kaplan-Meier rates? The assessment context and attribute tree exercises require explicit definitions of concepts and terms, and help stakeholders identify and resolve any differences in understanding.

1.3.3.2 Sharing an Assessment

There are numerous textual, tabular, and graphic means for presenting B-R assessments—many described in chapters throughout this book. While the data for B-R assessment in early development suffer from the limitations described earlier, to the extent that estimates of relative performance between the study treatment and a comparator (active or placebo) are available, clear depiction of the data and articulation of a B-R assessment are highly valuable. Even if actual comparative data are not available, the B-R framework can help us to define how big a treatment benefit will be required to ensure that, at least at a population level, benefits outweigh risks.

A fairly simple tool for B-R communication is an effects table[16] or a key B-R summary table [7] These are tables with all key benefit and harm attributes listed in separate rows, with values (rates, scores, time to events, etc.) provided for the study treatment, comparator(s), and potentially relative or absolute measures of differences between them (Table 1.1). An effects table typically includes attribute definitions, units, and ranges, while a key B-R summary table includes comparative measures with some degree of uncertainty. The key B-R summary table references the attribute definitions that

TABLE 1.1

Key Benefit-Risk Summary Table for the Value Tree in Figure 1.1

Endpoint	No. Events/10,000 Patient-Years		Risk Difference/10,000 Patient-Years (Study Drug—Comparator)	
	Study Drug	Comparator	N	95% CI
Cardiovascular death	400	423	−23	(−118, 72)
Nonfatal disabling stroke	44	66	−22	(−43, −1)
Nondisabling stroke	70	63	7	(−32, 46)
Nonfatal myocardial infarction (MI)	450	644	−194	(−311, −77)
Severe recurrent ischemia	586	620	−34	(−137, 69)
Nonfatal, nonstroke major bleeding	155	51	104	(66, 142)
Nonmajor clinically relevant bleeding	1960	1011	949	(801, 1097)

Note: All endpoints are on a 0–100% scale and shown as the number of events per 10,000 patient-years. All data are mock values.

are listed in a separate table. The intent is to have most critical information germane to a B-R decision in one location, ideally in units that allow for rapid evaluation, communication, and discussion of the relevant trade-offs. Despite the simplicity of these tables and the availability of the requisite data in full development, it remains unusual to find data on both benefits and harms presented in the same display.

Depending on the data available, more quantitative displays, such as forest plots, may also be valuable.[6,7,9,17]

1.4 Benefit-Risk in Early Development: Toward Full Development

1.4.1 Using B-R for Target Product Profiles and Other Core Documents

Many in pharmaceutical companies or regulatory agencies are familiar with the concept of a target product profile (TPP). The TPP is a summary of a drug development program's criteria for success, usually characterized in terms of concepts that are intended for the product label. TPPs serve as a communication medium between stakeholder groups within a sponsor organization and between a sponsor and health authorities.[18] In addition to benchmarks for safety and efficacy, TPPs often include nonregulatory considerations such as compliance, convenience, cost, access, and scheduling, all of which contribute to the broader picture of B-R. These additional consider-ations also speak to the fact that B-R assessments can be considered from the perspective of many stakeholders outside the sponsor organization, includ-ing providers, caregivers, and payers, in addition to patients and regulators. Figure 1.3 shows an example target product profile of an anticoagulant for acute coronary syndrome.

In addition to the TPP, product development teams within pharmaceuti-cal companies customarily maintain a compound core data sheet (CCDS), a cumulative summary of key information generated from studies about a product's clinical, nonclinical, and clinical pharmacology features. The TPP and CCDS are typically created during early clinical development and updated throughout the product's life cycle. Use of B-R methods such as those described in this chapter could integrate well with the TPP and CCDS to help decision makers, because the TPP and CCDS are distillations of data on outcomes and other considerations that are of greatest relevance to understanding the B-R profile of a product. Therefore, going through such exercises as building and maintaining an attribute tree and effects tables can help to populate information in the CCDS and TPP. Additionally, other more quantitative methods have been tested as a means of understand-ing the relative importance of different clinical outcomes for TPP develop-ment. These may be especially useful in settings where there is no available therapy or regulatory precedent.[12,13] In the case of the TPP, an attribute tree could be used to identify the key outcomes of interest or areas of uncer-tainty to guide the design of subsequent studies. In the case of the CCDS, annotated effects tables could provide a systematic and succinct format for documenting and describing information and referencing study reports. These could provide development teams with a concise means of synthe-sizing vast amounts of information from a variety of sources to build other documents created in early development, such as safety specification plans[19] and core risk management plans (cRMPs). Like the TPP and CCDS, these documents are used for guiding and describing a product's B-R profile dur-ing full development.

1.4.2 Using B-R in Gating Decisions

Stage gate decisions are used to determine which projects advance in a company's portfolio of projects. The advantages that we have discussed of structured approaches to B-R assessment could also be helpful in stage gate decisions. These include having a clear and shared understanding of the goals for the product's development (assessment or decision context), having a clear understanding of the endpoints/attributes needed for a positive B-R profile (attribute tree), and having tabular or graphic displays that allow the simultaneous depiction and communication of benefits and harms.

To make the best use of finite resources for developing a molecule, a com-pany will usually have criteria by which it evaluates the importance of a

	Efficacy	Safety/Tolerability	Value to Payer	Dose/Formulation	Indication
Study Drug for Acute Coronary Syndrome (Mock Target Comparator)					
Target comparator	• When added to standard of care, statistically significant relative risk reduction ≥ 20% in combined endpoints of CV death, MI, and stroke.	• Nonfatal major bleeding signifi antly higher (75%) than SOC • Nonmajor clinically relevant bleeding signifi antly higher (85%) than SOC	• Cost-effective compared to SOC	• 50 mg QD oral • Rapid onset of action • Higher loading dose	• Patients with acute coronary syndrome, both with medical management alone and with PCI
Study drug	• When added to standard of care, statistically significant relative risk reduction ≥ 20% in CV death. At least 10% relative risk reduction in MI or stroke.	• Nonfatal major bleeding no greater than 1% above that for SOC or target competitor • Overall major and nonmajor bleeding < 10% • No LFTs, no routine INR monitoring, no QT prolongation • No rebound after use • Fewer DDIs than both SOC and target comparator	• Cost-effective compared to both SOC and target competitor	• QD, oral, fixed dose • No leading dose needed • Rapid onset of action	• Patients with acute coronary syndrome, both with medical management alone and with PCI • Administration within 1 to 5 days of index event

FIGURE 1.3

Example target product profile of anticoagulant for acute coronary syndrome. All data for the target comparator and study drug are mock. CV = cardiovascular, DDI = drug-drug interaction, MI = myocardial infarction, PCI = percutaneous coronary intervention, SOC = standard of care. Assuming LFT = liver function test, INR = International Normalized Ratio (for blood clotting).

product in its portfolio. There are processes that make this process more quantitative and objective, although one could argue that the interpretation of such data in early development is ultimately a more qualitative and subjective assessment. Certain early development sciences, such as clinical pharmacology, have implemented model-based approaches to facilitate decision making that have even been embraced by health authorities.[20,21] While these generally apply to specific aspects of a product profile, such as the therapeutic window or dose-concentration-response time curve, these model-based approaches have been successfully applied to guide the design of full development studies using a "learn and confirm" approach.[22] Similarly, model-based approaches to B-R assessment could help us to align internal stage gate evaluations with potential regulatory concerns.

1.4.2.1 First-in-Human Decision

The decision to initiate studies in humans is one of the first points at which teams are able to use B-R type of considerations. Teams are tasked with evaluating the totality of available information to decide whether to file an application for human testing. The application undergoes a regulatory evaluation of the planned studies and supporting evidence.[23–25] Unlike in full development, this first B-R decision is based entirely on hypothetical benefits based on current understanding of the mechanism of action for the molecule identified in animal or in vitro tests, along with a nonclinical and in vitro safety evaluation. While nonclinical studies are performed in order to enable first-in-human studies, the team may have only just begun other studies that are important for painting a fuller picture of a product's B-R profile, such as nonclinical chronic toxicity studies.[26] B-R assessments made at this entry-into-human stage therefore rely on extrapolating attribute data solely from a limited number of nonclinical and in vitro observations. For some molecules, other than first-in-class compounds, inferences may be made from products with similar mechanistic activity that have reached this point in development or beyond.

1.4.2.2 Entry-into-Phase 2 Decision

An additional major B-R decision occurs after the initial human safety testing (end-of-phase 1). The data cumulative to this point are still too limited for quantitative B-R analysis, although the approaches described in Section 1.3 may be useful. Human studies reviewed at this stage will have ascertained the tolerability and acute toxic effects of the molecule, usually from single ascending and repeat dose trials, with human and nonhuman pharmacokinetic data. Scaling the drug exposure between animals and humans is also critical for examining whether toxicities seen in animals might occur at human therapeutic doses.

While a team's main goal at this second milestone is to rule out the presence of toxic effects on critical organ systems using toxicokinetic and pharmacokinetic data, comparing these toxic effects to expected benefits enables a more well-rounded decision. For diseases such as cancer, we might have the opportunity to establish proof of concept for efficacy while evaluating risks. For instance, in early development solid tumor oncology trials, typically conducted in patients with the target disease, one might evaluate observed or suspected treatment-related toxicities such as infusion reactions or class-related toxicities against signs of beneficial drug activity such as decreased glucose metabolism in tumor tissue.[27,28] However, compounds at this stage for many other diseases are generally tested only in healthy subjects, and evidence of benefit remains unknown at entry into phase 2. For these other diseases, the decision to proceed might be much more difficult since any clinical benefit at this stage is hypothetical.

1.4.2.3 End-of-Phase 2 Decision

Another major decision point that can incorporate B-R assessment occurs at the end of phase 2, the transition from early to full development. This assessment incorporates the proof of concept, dose-response, and longer-duration nonclinical studies from phase 2.[29] Subjects enrolled are patients with the targeted disease, rather than healthy subjects, so there is the potential to use proof of efficacy in B-R considerations.

End of phase 2 often provides a team its first opportunity to evaluate *comparative* B-R of the treatment. In phase 1 development, active controls are typically not used, so the assessment of B-R may have to be considered in absolute terms rather than relative to other treatment options. This leaves considerable uncertainty in the overall B-R profile. Depending on the disease area, phase 2 may provide an initial understanding of comparative B-R: how the benefits and risks of the product compare to those of an alternative treatment.[30] The choice of what doses to bring into phase 3 may also be informed by a B-R perspective. For example, the adverse event rates might be increased at higher doses without an equivalent increase in benefits. Similarly, use of an active comparator in phase 2 can provide information on how the B-R profile of the new compound compares to existing therapies.

Because of the huge resources required for late-stage product development, pharmaceutical developers face great selective pressure to boost productivity by eliminating products with unfavorable B-R profiles at an early stage.[31] However, because early development is characterized by shorter timelines and higher failure rates than late development, teams need to reach decisions about B-R quickly. To these ends, the robust and transparent nature of good B-R assessment methods and their application can provide a structured approach to perform the critical B-R assessments that can greatly assist in portfolio decisions.

1.5 Conclusions

The B-R concepts and methods proposed in this chapter for early development have been tested largely on molecules in either the late-stage or post-licensing phases of product development. Given the value propositions for using B-R methods in early development, and given that early development is the foundation upon which full development rests, it is actually surprising that B-R methods have not been more formally used routinely in early development. The irony is that since regulatory review for licensing occurs several years after early development ends, there is less pressure in early development to identify and employ a validated B-R methodology. The longer time frame would provide an opportunity to test different ways of setting up the assessment when the costs for such testing are relatively low. Because of the shorter timelines in early development, there is an opportunity to more rapidly explore new measurements and endpoints to use in B-R assessments, which could then be added as exploratory or additional formal endpoints to early development studies and used for preliminary quantitative assessments.

While we have highlighted several of the key early development milestone decisions where B-R assessment may be helpful, the reality of drug development is that data are continuously being generated and evaluated by product teams. Decision or assessment contexts, attribute trees, and means to communicate the assessment will all change through the course of development. In all stages, attribute trees and other B-R tools can facilitate our real-time interpretation and corresponding decision making and future planning as they relate to changing data and changing objectives. Since the greater speed of early development can come at the cost of lessening the planning time available to a clinical development team, the structured approaches in a typical B-R framework can help teams better prepare for internal stage gate meetings, thereby guiding the decisions that lead a molecule through full development.

References

1. *Periodic benefit-risk evaluation report (PBRER) E2C (R2)*. 2012. International Conference on Harmonization.
2. Noel, R., Hermann, R., Levitan, B., Watson, D.J., and Van Goor, K. 2012. Application of the Benefit-Risk Action Team (BRAT) Framework in pharmaceutical R&D: Results from a pilot program. *Drug Information Journal* 46: 736–743.
3. *PDUFA reauthorization performance goals and procedures*. 2011. FDA. http://www.fda.gov/downloads/forindustry/userfees/prescriptiondruguserfee/ucm270412.pdf.

4. European Medicines Agency Benefit-Risk Methodology Project, EMA/213482/ 2010. 2010.
5. Guidance for industry and Food and Drug Administration staff: Factors to consider when making benefit-risk determinations in medical device premarket approval and de novo classifications. 2012. FDA CDRH. http://www.fda.gov/MedicalDevices/DeviceRegulationandGuidance/GuidanceDocuments/ucm267829.htm.
6. Coplan, P., Noel, R., Levitan, B., Ferguson, J., and Mussen, F. 2011. Development of a framework for enhancing the transparency, reproducibility and communication of the benefit-risk balance of medicines. *Clinical Pharmacology and Therapeutics* 89: 312–315.
7. Levitan, B., Andrews, E., Gilsenan, A., Ferguson, J., Noel, R., Coplan, P., and Mussen, F. 2011. Application of the BRAT Framework to case studies: Observations and insights. *Clinical Pharmacology and Therapeutics* 89: 217–224.
8. Walker, S., McAuslane, N., and Liberti, L. 2011. Developing a common benefit-risk assessment methodology for medicines—A progress report. *Scrip Regulatory Affairs*.
9. Innovative Medicines Initiative PROTECT Project. 2012. http://www.imi-protect.eu/index.shtml.
10. Frey, P. *Benefit-risk considerations in CDER: Development of a qualitative framework.* 2012. http://www.fda.gov/downloads/AboutFDA/CentersOffices/OfficeofMedicalProductsandTobacco/CDER/UCM317788.pdf.
11. Dodgson, J., Spackman, M., Pearman, A., and Phillips, L. 2009. *Multi-criteria analysis: A manual.* London: Department for Communities and Local Government.
12. Cross, J. 2013. The relevance of patient treatment preferences to endpoint selection. CNS Summit, Boca Raton, FL.
13. Bridges, J., Kinter, T., Kidane, L., Heinzen, R., and McCormick, C. 2008. Things are looking up since we started listening to patients: Trends in the application of conjoint analysis in health 1982–2007. *Patient* 1: 273–282.
14. Keeney, R.L. 2002. Common mistakes in making value trade-offs. *Operations Research* 50: 935–945.
15. Belton, V., and Stewart, T. 2001. *Multiple criteria decision analysis: An integrated approach.* Boston: Kluwer.
16. Zafiropoulos, N., Phillips, L., and Pignatti, F. 2012. Evaluating benefit-risk: An agency perspective. *Regulatory Rapporteur* 9: 5–8.
17. Levitan, B. 2011. A concise display of multiple end points for benefit-risk assessment. *Clinical Pharmacology and Therapeutics* 89: 56–59.
18. *Guidance for industry and review staff: Target product profile—A strategic development process tool.* 2007. FDA CDER. http://www.fda.gov/downloads/Drugs/GuidanceComplianceRegulatoryInformation/Guidances/ucm080593.pdf.
19. International Conference on Harmonisation. *ICH tripartite guideline E2E pharmacovigilance planning.* http://www.ich.org/fileadmin/Public_Web_Site/ICH_Products/Guidelines/Efficacy/E2E/Step4/E2E_Guideline.pdf.
20. Peck, C., et al. 1992. Opportunities for integration of pharmacokinetics, pharmacodynamics, and toxicokinetics in rational drug development. *Journal of Pharmaceutical Sciences* 81: 605–610.
21. Woodcock, J., and Woosley, R. 2008. The FDA critical path initiative and its influence on new drug development. *Annual Review of Medicine* 59: 1–12.

22. Sheiner, L. 1997. Learning versus confirming in clinical drug development. *Clinical Pharmacology and Therapeutics* 61: 275–291.

23. Part 312 Code of Federal Regulations, Title 21. FDA. http://www.accessdata. fda.gov/scripts/cdrh/cfdocs/cfcfr/cfrsearch.cfm?cfrpart=312.

24. Guidance for industry content and format of investigational new drug applications (INDs) for phase 1 studies of drugs, including well-characterized, therapeutic, biotechnology-derived products. 1995. FDA CDER, CBER. http://www.fda.gov/downloads/Drugs/GuidanceComplianceRegulatoryInformation/Guidances/ucm071597.pdf.

25. *EU clinical trial directive, clinical trials directive (Directive 2001/20/EC of 4 April 2001).* http://ec.europa.eu/health/files/eudralex/vol-1/dir_2001_20/dir_2001_20_en.pdf.

26. Guidance for industry: M3(R2) nonclinical safety studies for the conduct of human clinical trials and marketing authorization for pharmaceuticals. 2010. FDA CDER, CBER. http://www.fda.gov/downloads/Drugs/GuidanceComplianceRegulatoryInformation/Guidances/ucm073246.pdf.

27. Kelloff, G., et al. 2005. Progress and promise of FDG-PET imaging for cancer patient management and oncologic drug development. *Clinical Cancer Research* 11: 2785–2808.

28. Dillman, R.O. 1999. Infusion reactions associated with the therapeutic use of monoclonal antibodies in the treatment of malignancy. *Cancer Metastasis Review* 18: 465–471.

29. *Guidance for industry: Exposure-response relationships—Study design, data analysis, and regulatory applications.* 2003. FDA CDER, CBER. http://www.fda.gov/downloads/Drugs/GuidanceComplianceRegulatoryInformation/Guidances/ucm072109.pdf.

30. International Conference on Harmonisation. 2000. *Choice of control group and related issues in clinical trials E10 guidelines.* http://www.ich.org/fileadmin/Public_Web_Site/ICH_Products/Guidelines/Efficacy/E10/Step4/E10_Guideline.pdf.

31. Paul, S., Mytelka, D., Dunwiddie, C., Persinger, C., Munos, B., Lindborg, S., and Schacht, A. 2010. How to improve R&D productivity: The pharmaceutical industry's grand challenge. *Nature Reviews Drug Discovery* 9: 203–214.

Section II

Full Clinical Development

Introduction

The door to the Full Clinical Development gallery is well used, and the space beyond is expansive and inviting. The walls are adorned with several beautifully rendered works, carefully arranged to tell a story. It is a story of inquiry, construction, deliberation, and communication.

The first paintings we encounter depict debate and concern, men and women engaged in discussions that appear to both elate and dishearten. The images signify the questions and concerns that drive inquiry, and the discourse and frameworks that enable the articulation of questions, as well as suggesting the paths to answering those questions.

Following these paintings, we encounter depictions of structures and devices, constructs and machines, both real and imagined. The stuff of understanding. The imaginings and workings of problem solvers.

Last is a single painting of a woman reading a letter. Her face is inscrutable, as likely enlightened as mystified. Upon closer examination, we recognize her as one of the women in the first painting we viewed, and we catch glimpses of the etchings of structures and machines of the later paintings in the letter she holds.

2

Key Questions, Issues, and Challenges in Benefit-Risk Assessment in Full Clinical Development

John Ferguson and Marilyn Metcalf

CONTENTS

> If I had one hour to save the world, I would spend 55 minutes defining the problem and only five minutes finding the solution.
>
> **—Albert Einstein**

We use the terms *benefits* and *risk* to describe the effects of medical interventions. More correctly, they are *potential* benefits and *potential* harms, because medicine can potentially change people's lives for better or worse. For any given individual, we do not know what the specific outcomes will be. The goal for individuals and organizations that develop, approve, prescribe, and market drugs, biologics, vaccines, and devices is to maximize the probability that the user will enjoy meaningful benefits without sustaining undue harm. Systematically assessing benefit-risk (B-R) provides a means for defining this goal and measuring the effects of treatment choices against it.

We define and understand benefits by articulating what is important about the outcomes we seek for patients. For example, if we seek a medical intervention that improves physical mobility or cognitive functioning, it is because it is important for the patient to be able to live as independently and normally as possible. Assessing benefit involves rationally identifying and defining clinically relevant outcomes, biomarkers, or other surrogates that are considered favorable effects for the patient. Along with identifying and defining the outcomes of interest, it is important to know their intensity, duration, and the attendant uncertainty that surrounds them in order to paint a more complete picture of the impact on the patient.

Benefits move beyond efficacy and effectiveness. Efficacy refers to product performance under ideal, highly controlled circumstances. As such, efficacy studies often provide good answers to the wrong questions: results may be internally valid but lack external validity, and as a result, do not easily generalize to ordinary clinical practice.[1,2] Effectiveness adds real-world relevance to efficacy results. It refers to product performance under conditions of usual use. However, there is a price to pay for increased relevance—reduced internal validity due to reduced control of bias. Benefit includes an understanding of the endpoints' relevance and perceived value, from the point of view of the patient, the clinician, or the regulator.

The fundamental issue is similar for risks, which extend beyond study safety information. Safety data are collected during clinical trials, observational studies, and through spontaneous reports. Risks, or more specifically, potential harms, are clinically relevant outcomes, biomarkers, or other surrogates that are considered unfavorable effects of medical interventions. We understand harms not only in their occurrence, but also in how we manage the events when they occur in order to minimize their impact. To assess the risks associated with harms, we need to know the frequency, severity, duration, predictability, monitorability, and reversibility of those harms. If we can find a harm early enough to mitigate or prevent a negative outcome, it changes the way we view the risk associated with that harm.

Historically, healthcare decisions have depended on the mostly disconnected assessment of proxies for benefit and risk (i.e., efficacy and safety). The B-R trade-off examines how the favorable effects of a treatment compare to its unfavorable effects. Drug researchers and manufacturers, regulators, payers, healthcare providers, and patients all need dependable information for decision making. Structured qualitative and quantitative B-R assessment is increasingly important in helping determine whether the favorable effects of a treatment sufficiently outweigh the unfavorable ones to justify its use.[3–6]

But as the quote from Einstein reminds us, we need to be asking the right question to find the right answer for each of these stakeholders. The B-R context matters. First, we need to ask whether any treatment is better than no treatment. This question often comes up in discussions around so-called lifestyle diseases or conditions, which may not be considered by some stakeholders to have serious health outcomes, while other stakeholders may consider their impact to be devastating on the lives of patients. Second, there are conditions that may have more serious associated morbidities but are not immediately life threatening. If treatment includes rare but serious and potentially fatal side effects, views may differ regarding the acceptability of the treatment. Even with the treatment of a life-threatening condition, there may be questions about how much benefit is enough. If, on average, life is extended by a small amount, is that because a few patients benefit greatly while others do not experience the benefit, and we need to understand better the appropriate patient population? Or do most patients see a modest gain, and must we ask how important is that gain to them, given the side effects they may experience?

The context for comparing benefit to risk may change again when one is using a medicine prophylactically versus using it as an intervention for an existing disease or condition. Likewise, the perceived implications of short-term versus long-term risks or gains may vary with the nature of the disease or condition, as well as with the circumstances of the patient and his or her preferences. These questions about the value of outcomes ultimately push us to examine the fundamental goals of healthcare, which may be influenced by cultural context and patient population, as well as by perspective.

Framing benefits and risks in a manner that addresses the complexity of the contexts in which medicine is used increases the likelihood of achieving a rational, internally consistent adjudication of the B-R trade-offs. A framing exercise follows a well-defined process that identifies the question(s), identifies the benefits and risks and how they are to be measured, and incorporates and highlights the evidence for favorable and unfavorable effects. A qualitative or semiquantitative approach to B-R evaluation may stop here, or a more rigorous quantitative assessment may be needed for clarity. Empirically, the importance of proper framing increases in direct proportion to the complexity of the B-R problem at hand.[7]

As noted above, in addition to context, there are also different perspectives on the value of treatment. Weighting is a structured way to capture this diversity of thinking, although not an answer in and of itself. Weights, or preferences, can be difficult to collect and context specific, but there are established methods to elicit them. A skilled practitioner can facilitate an insightful discussion and lay the foundation for an analysis that represents diverse views. For example, it is not surprising that both the psoriatic patient and his or her physician find it hard to understand how an improvement in the standard Psoriasis Area Severity Index (PASI) score should be weighed against the possibility of developing lymphoma as a consequence of taking an antipsoriatic treatment. On the other hand, they may arguably be able to trade off the likelihood of developing lymphoma against the more meaningful, albeit more difficult to measure, likelihood of their skin condition improving to the point where they can go to the beach without feeling self-conscious.

Acknowledgment of different perspectives provides the basis for transparent communications among groups that support different weights. Demonstrating an individual patient's perspective and an understanding of how and why it may differ from that of a regulator's public health perspective, or from a healthcare provider's clinical perspective, is vital to the transparency of healthcare decision making. It is this "meta-dialogue" that allows us to incorporate the different perspectives in decision making. An estimation of the balance of value-weighted trade-offs can allow key stakeholders, including healthcare providers and patients, to make more informed therapeutic decisions. Data limitations and collection techniques notwithstanding, the decision-making process can be improved by careful structuring, quantification, and standardization, as well as by increased transparency and broader stakeholder input (particularly as the process relates to end user values).

Recently, there has been some mention in regulatory meetings and publications regarding the inclusion of B-R information in risk communications.[8,9] At the same time, simple disclosure is not transparency and does not guarantee understanding. Interpretation and clarification are necessary, and implicitly incorporate value judgments that need to be acknowledged. Assuming the communicators reach understanding, that understanding does not automatically ensure agreement. It is likely that differences will remain, but the goal is to have a functional system that works well with those differences. Each approved medicine is not right for every patient, nor will every patient be prescribed every medicine or choose to take it. The greater challenges come when medicines are not approved or are not included on a formulary, or the prescriber refuses to write the prescription, and the patient has no choice. The role of the patient versus those of regulators, payers, and prescribers is still being determined.[10] While current methods for analysis and weighting are not perfect, they are adequate as a starting point. The B-R field will move forward, not just with improved methodologies, but with political will that provides clear policies for the exchange and use of B-R information.[11] Healthcare organizations are exploring the use of B-R in decision making, and the stage is being set for shared standards of assessment. B-R information is already required in some regulatory documents, and may be used increasingly in risk communications. The future will bring these efforts together in a more cohesive fashion, to create a forum for more formal use of B-R information to fulfill its promise of improved patient outcomes.

References

1. Fletcher, R., Fletcher, S., and Wagner, E. 1996. *Clinical epidemiology. The essentials.* 3rd ed. Baltimore: Williams and Wilkins.
2. Feinstein, A. 1983. An additional basic science for clinical medicine. II. The limitations of randomized trials. *Annals of Internal Medicine* 99: 544–550.
3. *The future of drug safety: Promoting and protecting the health of the public.* Institute of Medicine. September 22, 2006.
4. Philips, L. 2011. *Benefit-risk methodology project. Work package 3 report: Field tests.* EMA/718294/2011.
5. Mussen, F., Salek, S., and Walker, S. 2009. *Benefit-risk appraisal of medicines: A systematic approach to decision-making.* Chichester: Wiley-Blackwell.
6. European Medicines Agency. Benefit-risk methodology project. Work package 4 report: Benefit-risk tools and processes. EMA/297405/2012—Revision 1. 2012.
7. Moore, R., Derry, S., McQuay, H., and Paling, J. 2008. What do we know about communicating risk? A brief review and suggestion for contextualising serious, but rare, risk, and the example of cox-2 selective and non-selective NSAIDs. *Arthritis Research and Therapy* 10 : R20. http://arthritis-research.com/content/10/1/R20.

8. http://www.fda.gov/AdvisoryCommittees/CommitteesMeetingMaterials/RiskCommunicationAdvisoryCommittee/default.htm.

9. http://www.ema.europa.eu/docs/en_GB/document_library/Report/2011/05/WC500106865.pdf.

10. Eichler, H.-G., Abadie, E., Baker, M., and Rasi, G. 2012. Fifty years after thalidomide; what role for drug regulators? *British Journal of Clinical Pharmacology* 74: 731–733.

11. Noel, R. 2012. Weighing and valuing evidence in drug development and regulatory decision making: A situational analysis. Presented at Center for Innovation in Regulatory Science, Philadelphia, December 13, 2012.

3

The Clinical Aspects of Benefit and Risk

Marilyn Metcalf

CONTENTS

We have seen already that when we try to define and measure benefit, we are moving beyond the biological measures intended to test whether a drug works (efficacy/effectiveness). With benefit, we are interpreting *what* we want a treatment to do for patients when it works (the clinical outcomes that result from efficacy or effectiveness) and *why* we want it to work (what is important about the outcomes). This understanding of benefit is fundamental to healthcare. It is how we find the patients', healthcare providers', drug makers', and regulators' perspectives. Likewise, for risk we need to understand the consequences to the patient and how to manage any adverse events (AEs) that might occur. Processes for evaluating benefit versus risk seek to determine whether a treatment, on balance, contributes to a healthy state for a population and for individual patients. To make that determination, we must ask incisive questions, in a clear manner, and in a way that allows for unambiguous responses. At the same time, we must allow for the different perspectives that may define health and benefit-risk (B-R) differently, and therefore may give unambiguously different responses to the questions.

The way to identify those kinds of questions and account for differing perspectives is through *framing*. Framing is a utilitarian term that combines the focus and boundary of a picture frame with the structural skeleton that a house frame brings to mind. That is, in framing the B-R questions around a medicine, we focus on the key content and then build the evidence accordingly. As we move from asking the questions to finding answers in a structured way, we shift to using a *framework*. Key elements for a B-R framework include governance, measurement, graphics (optional but very helpful), clinical relevance, and pharmacovigilance. All of these rudiments serve as a backdrop for an informed discussion of benefit and risk. Most published and presented B-R frameworks[1-5] are based upon decision analytic techniques that help define goals, options, and trade-offs between alternatives. They do not "make the decision." Ultimately, they comprise evidence in what must still be a judgment call by human decision makers.

The elements in these frameworks fit together to provide a broad picture of a medicine for the decision makers who evaluate it throughout its life cycle.

1. **Governance:** B-R assessments are multifaceted. At heart, they are designed to share fundamentally important information with a wide variety of people who are interested for a number of different reasons. Finding common language and ways of delivering and understanding this information requires some organizational structure and governance to provide a consistent set of standards from which to work. Within those standards, there needs to be a great deal of flexibility, but at a minimum, there is a growing ethos that medicines and medical devices must undergo B-R assessments as part of their development, regulatory review, and ongoing availability to patients.

   a. When a medicine or device is in development, the responsible project team will report to internal and external review boards as the drug progresses. The B-R balance must remain favorable for the drug to progress through each stage. When the project development teams present B-R, they describe the intended use of the medicine/device, the patient population, the (perhaps unmet) medical need being addressed, and some comparison to alternative treatments (that may include no treatment or placebo). If surrogates are used for clinical benefit and risk (e.g., laboratory values used to monitor chronic diseases), teams should describe how these surrogates are expected to translate to clinical benefit or discharge of risk. They need to place as much emphasis on the level of evidence for risk as they place on evidence for benefit, so that each is seen in an appropriate context with the other.

   b. Medical industry regulators are beginning to ask for more explicit presentations of B-R as part of the clinical description of medicines in initial regulatory applications and in updated information reports. The agencies are also exploring the application of B-R frameworks in their own decision making around medicines and devices. These reviews ask similar questions to those described above, around medical need, other available options, evidence that demonstrates how the medicine will help the patient, and what unintended harms may occur. If the drug or device is approved, the sponsor company will continue to collect information over the product's life and provide that information to regulators, who will in turn continue to assess whether there is evidence that the product should remain available to patients. These are important touch points at multiple times in a medical treatment's life cycle, where vital questions about clinical benefit and AEs must be asked and answered.

2. **Measurement:** Frameworks require that specific benefits and risks be identified and defined with regard to how they are measured. Generally, a group of experts (e.g., development team members, regulators, healthcare providers, patients) follow a well-defined process to identify the outcomes that are of greatest importance in describing a drug's benefits and risks. These experts are asked to consider the factors above, such as whether the medicine meets an unmet need for patients or how the effects of the medicine are distinct from those of other treatments. The measures for the outcomes of interest may be ordinal (e.g., more than/less than, high/medium/low) or more numeric. The outcomes should be clearly defined. A benefit like "improved mobility" should be more clearly defined with one or more measurable outcomes, such as "ability to walk 10 yards/meters," and measured using an appropriate functionality scale or instrument. A risk of something like a "liver event" likewise needs to be defined as to whether it means simple elevation of liver enzymes or represents criteria for something more serious that may suggest liver failure.

These explicit definitions are an exercise in objectivity. The process of defining a model that identifies and measures benefits and risks adds clarity and insight into how the participants are thinking about the issues. The exercise provides shared understanding for those building the model as well as clear ways of communicating their thoughts to others. While almost any B-R assessment will require some type of quantitative measure, this does not mean that all assessments require lengthy analyses. It does, however, push for a certain amount of transparent thinking to define not just benefit and risk, but *how much* of each is present. Risk differences or relative risks, based on proportions of patients experiencing the benefits and the risks of a medicine versus an alternative treatment, have been used. Even this simple approach can be quite revealing in understanding the magnitude of the population who are affected favorably and unfavorably by a treatment. Further exploration may reveal additional insights regarding whether certain subgroups of patients experience mostly favorable effects, mostly unfavorable effects, multiple AEs, high or low levels of benefit or risk, both greater benefit and more risk, or neither benefit nor risk. At a population level, it is important to know what tends to happen across a range of patients; at an individual level, guidance is needed regarding whether there are signs that a patient may experience greater benefit or greater risk, and whether there are early signs that may help that patient avoid more serious complications given a harmful event.

3. **Graphics:** The intent of a B-R graph or chart is to provide a transparent visualization that captures in a single place key benefits and risks, preferably along with their uncertainty. Different visualizations are appropriate for different types of data. If benefit and risk can be measured in the same manner, they can often be represented in the same graphic. If not, they may need to be included in juxtaposed graphics. B-R visualization is an emerging area that will continue to develop, although there already exist a number of excellent references for creating appropriate graphics from a statistical and visual point of view.[6,7] The goal of a benefit and risk visualization is to provide comparative visual information while maintaining the integrity of the data and their implications for patient well-being. Well-made graphs can provide an overview clearly and quickly. As with other areas of B-R evaluation, graphics are designed to inform discussion and decision making. Many of the graphs made to date display key benefits and key risks, along with the uncertainty around them (e.g., 95% confidence intervals), side by side, usually against a comparator. Graphs that can accommodate continuous data, display levels of intensity for benefit or severity for risk, or help demonstrate the mixture of benefit and risk experienced by different types of patients may help to enhance our understanding as well.[8]

4. **Clinical relevance:** Per the above, measures of benefit and risk need to be both objective and clinically meaningful. To define a benefit, we must understand why, for example, a "20% change from baseline" matters. Implicit assumptions in different disease areas need to be made explicit. Acceptable trade-offs between benefits and risks may be quite different in assessments for a treatment for a life-threatening condition versus a chronic but nonfatal condition versus disease prevention in a healthy population. These are areas in which one's perspective plays a large role. With a disease as diverse as cancer, there may be broad ranges of patients who are experiencing risks from family history, risk of recurrence, isolated disease and recovery, chronic disease, metastatic disease, and long-term survival. Within these groups of patients, individuals will have individualized views of themselves, their experiences, their desires for their health, and the ways that they think about the trade-offs among treatments. We will be looking in the chapters that follow at methods for capturing diverse viewpoints and values.

5. **Pharmacovigilance:** As mentioned above, data about drugs and devices are collected throughout a product's life cycle. Risks that are of special concern are addressed by drug and device companies using specific plans agreed to with regulators as part of the approval process. Increasingly, these plans are explicitly requiring that B-R balances be monitored. Characteristics of risks in these plans

include monitorability, meaning an indication of whether a risk can be anticipated and tracked, rather than appearing full blown in a patient without warning. If a risk can be monitored (e.g., with regular blood tests or other health checks), treatments can be stopped or changed before a serious event occurs, and there is a greater level of confidence that serious harm can be avoided. Reversibility is another important characteristic of some risks. If an AE will stop and disappear once a patient stops treatment, the patient may be able to stop a treatment and restart it after a rest period (dose interruption) or switch to another therapy before accruing serious harm. Reversibility can reduce risks for patients who try a therapy only to find that it is not for them. On the other side of the coin, duration of benefit is an important part of the ongoing B-R balance that must be demonstrated. Some patients build up tolerances for therapies that may necessitate increased dosages, for example. It is important to note whether such patients tend to maintain a level of benefit, whether increased dose carries increased risks, and whether those risks can be monitored and potential harms reversed. Understanding the course of both the favorable and unfavorable effects of a treatment, particularly a chronic one, helps better characterize the experience of patients.

Subgroups of patients like those described above may experience different benefits and risks, or may approach balancing benefit and risk differently. Traditionally we have been accustomed to thinking of demographic subgroups or genetic subgroups of patients for targeted therapies, but as noted at the FDA's Inaugural Patient Network Meeting,[4] patients may more easily coalesce with others who have similar life experiences or goals. Below are some examples of priorities for therapy for different subgroups of patients. This list is not comprehensive, but is meant to help compare and contrast some of the B-R trade-offs.

Disease prevention: An otherwise healthy person considering a prophylactic medicine for himself or herself or a child is likely to be primarily concerned with a product's safety. His or her interest would be in having a vaccine or medicine that provides durable benefit with little risk and minimal invasiveness in its administration. The safety threshold for these patients is likely to be quite strict, although how it balances against benefit may vary with both the seriousness of the disease or condition to be prevented and the patient's perception of his or her susceptibility or that of his or her child.

From a public health perspective for products like vaccines, regulators and health authorities are interested in the benefits of "herd immunity" (i.e., ensuring that as many people as possible are protected to decrease the likelihood of widespread infection). This approach to disease prevention may tempt individuals to avoid participation, in the hopes that others will take any risk associated with prophylaxis, while the unvaccinated might still receive the benefit of a low infection rate. Other individuals may believe that protection from infection is more assured only by taking the vaccination. Safety is of special concern, as these types of products are meant to be used widely and are often recommended for vulnerable populations. B-R trade-offs will tend to be based on projections of reduced rates of infection and consequent sequelae versus estimated rates of potential AEs associated with prophylaxis.[9]

Chronic treatment: Similar in some ways to prevention of disease is an intervention to keep a disease at bay. Patients with chronic conditions would be expected to need treatments that are durable and cause minimal or tolerable AEs. Their needs might in part also depend upon their desired activity level.[10] For example, patients accustomed to more vigorous lifestyles might prioritize benefits of treatment options differently to preserve their sense of self, versus others with more sedentary lifestyles who might see benefits differently. Patients trying to prevent recurrence or relapse of a disease might have more knowledge of their chances for recurrence, which may give them a different perspective on benefit versus a potentially unknown risk of an AE. In these circumstances, the considerations may be more around projections and perceptions of life expectancy and quality of life with treatment versus without treatment.

Acute life saving: Patients in emergency situations (e.g., cardiovascular events) would be expected to make B-R trade-offs quite different from those described above. In acute situations, a greater emphasis would be expected on treatments that are immediate and highly effective. It is desirable that any AEs would be manageable, but life-saving treatments are likely to be of short duration, and the first goal is to preserve life that is imminently at risk. Patients and surrogates may not be in a position to have a choice at all in the crucial moments available to them for action. Healthcare providers must act quickly, based on often limited information, and have a management plan afterwards to follow up on AEs.

Curative: Patients with potentially fatal diseases or conditions would likewise be expected to focus on the efficacy of a product. There is a desire for targeted therapy for many diseases and conditions, as the choices of first- and second-line treatments can be crucial to the patient's survival. Durability would also be important in maintaining clearance of disease or remission of condition. In these circumstances, one might find oneself comparing immediate survival benefits with both short- and long-term risks.

End of life: Patients who have reached the final stages of a disease or condition may or may not be continuing to receive treatments. They may be seeking care that will make them comfortable and cause minimal AEs. They may have goals for things they would like to accomplish or people they would like to see and may be looking for treatments that will best allow them to achieve those goals. Their care may depend upon whether they would like to be, or can be, in a hospital, hospice, or at home. Increasingly, there are also considerations within healthcare payer systems and even among healthcare providers regarding the costs associated with treatment in the final stages of a disease.[11] Patients and their families may have very difficult choices to make concerning how to approach care at the end of life. At the population level, payer systems are making choices regarding what constitutes a benefit for patients in these circumstances and what limits should be placed around payment for them. Risks may be viewed in the context of whether they detract from the patient's goals.

If therapeutic requirements are viewed in this way, as needing to cater to circumstances and treatment goals, then there may be an opportunity for shared standards for treatments or treatment types. Therapies might be judged on how well they meet those standards, and individual patients might receive information that addresses key therapeutic characteristics so that they and their healthcare providers are better informed to make their decisions. One way of approaching these standards would be to answer the questions shown in Figure 3.1. As noted, the answers to these questions might have different priorities among different individuals or groups, but if we strive to have this information available for medicines, it will enable decision makers to apply it to their needs.

Readers familiar with health economics, cost-effectiveness analysis, and health technology assessments may note that measures such as cost per quality-adjusted life year (QALY) or efficiency measures in financial analyses are not usually assessed in a strict B-R analysis. B-R evaluation emphasizes a treatment's contribution to the patient's clinical course or experience of disease without using costs or cost savings as a factor. Cost-effectiveness

	Benefits	**Risks**
Intensity	How good are they?	How severe are they?
Time	How soon do they happen? How long do they last?	How soon do they happen? How long do they last?
Probability	Do they only happen for some people?	Can they be avoided? If no, can they be managed?

FIGURE 3.1
Key benefit-risk questions to answer.

might measure reduced numbers of days in hospital and the consequent money saved, while B-R might measure improved efficacy or potency that reduces number of days of treatment needed, with a close look at whether increased potency might increase risk of AEs or decreased duration of dosing might reduce risk of AEs. The measures should not be contradictory, and can inform one another, but serve different purposes for different audiences.

The separation between the clinical and the financial may be in part due to the diversity of payment programs for medicines on a global scale. In countries with nationalized payment plans and drug formularies, there may be more alignment between assessing the clinical benefits and the cost benefits of drugs. In other countries with more private payer systems, the decisions may still be quite distinct, depending upon the responsibilities of the decision makers (although as noted above, alignment may be growing). In any environment, clinical and regulatory decision makers focus on the outcomes to the patient's health, and are among the primary stakeholders for B-R evaluation. Increasingly, patients and the general public are included as stakeholders as well. They are participants in generating B-R data and assessing the information that is important to include in B-R communications. Patients participate in decision making for themselves and in making their voices heard among the other B-R stakeholders. Furthermore, it is important to understand the context from which a patient speaks: as noted, patients can have diverse needs, even within the same disease.

Patients are both the potential beneficiaries of more effective treatments and, at the same time, those who bear the associated risks. Regulatory authorities recently have acknowledged the need for quantitative assessments of patients' risk tolerance to support decisions involving B-R trade-offs. Chapter 4 introduces readers to best-practice methods for quantifying how much risk of adverse outcomes is acceptable to patients for treatments that offer improvements in efficacy. B-R thresholds can be estimated using accepted stated preference techniques such as discrete choice experiments, and threshold estimates can be used to calculate metrics such as maximum acceptable risk, minimum acceptable efficacy, net efficacy benefit, and net safety benefit. Policy-relevant B-R preference studies must satisfy a number of conditions, including conceptual validity, use of best-practice risk communication and survey methods, clear correspondence to clinical trial endpoints, recruitment of an appropriate patient sample, and correct statistical analysis of the data. The authors employ an empirical example to illustrate the steps required to complete a discrete choice experiment study and interpret the resulting preference estimates.

If we move from a patient-centric measurement to analyzing B-R at a group level, we find that discussion, debate, and voting are currently used by pharmaceutical company managers and regulators to establish the B-R balance of medicinal products. The search for more consistent, transparent, and communicable methods is motivating research that has established proof of principle of structured and quantitative methods. Chapter 5 presents necessary and sufficient conditions for rigorous and scientifically valid approaches to quantifying B-R. It details a case study that illustrates how a group of key players working in a facilitated workshop constructed a quantitative B-R model that captured diverse sources of data, clinical judgment about the relevance of the data, and assessments of trade-offs between the favorable and unfavorable consequences of treatment. Exploration of the model fostered understanding and commitment among participants about the key issues, thereby enabling effective decision making. The chapter ends with some speculation about why explicit, quantitative methods are found challenging, and what the future might hold for a shift from implicit to explicit, and qualitative to quantitative assessment of the B-R balance of drugs.

References

1. Coplan, P., Noel, R., Levitan, B., Ferguson, J., and Mussen, F. 2011. Development of a framework for enhancing the transparency, reproducibility and communication of the benefit-risk balance of medicines. *Clinical Pharmacology and Therapeutics* 89: 312–315.
2. European Medicines Agency Benefit-Risk Methodology Project. http://www.ema.europa.eu/ema/index.jsp?curl=pages/special_topics/document_listing/document_listing_000314.jsp&mid=WC0b01ac0580223ed6.
3. Innovative Medicines Initiative. http://www.imi.europa.eu/content/home#&panel1-4.
4. U.S. Food and Drug Administration. http://www.fda.gov/ForConsumers/ByAudience/ForPatientAdvocates/ucm298136.htm.
5. Centre for Innovation in Regulatory Science. http://cirsci.org/UMBRA.
6. Duke, S. 2012. *Graphics design and the biostatistician. Biopharmaceutical report*, 4–9. Vol. 19, no. 2. American Statistical Association.
7. Tufte, E. 2001. *The visual display of quantitative information*. 2nd ed. Graphics Press, Cheshire, CT, USA.
8. Norton, J. 2011. A longitudinal model and graphic for benefit-risk analysis, with case study. *Drug Information Journal* 45: 741–747.
9. Richardson, V., et al. 2010. Effect of rotavirus vaccination on death from childhood diarrhea in Mexico. *New England Journal of Medicine* 362: 299–305.
10. Kachuck, N. 2011. When neurologist and patient disagree on reasonable risk: New challenges in prescribing for patients with multiple sclerosis. *Neuropsychiatric Disease and Treatment* 7: 197–208.
11. Bach, P., Saltz, L., and Wittes, R. 2012. Op-ed contributor: In cancer care, cost matters. *New York Times*, October 14, 2012.

4

Quantifying Patient Preferences to Inform Benefit-Risk Evaluations

F. Reed Johnson, A. Brett Hauber, and Jing Zhang

CONTENTS

4.1 A Conceptual Framework

4.1.1 Benefit-Risk Evidence and the Weighting Problem

Pharmaceutical benefit-risk (B-R) evaluations pose many measurement challenges, including multiple potential efficacy and safety endpoints; large uncertainty about the likelihood, causality, reversibility, and latency of low-frequency adverse events; and heterogeneity of effects among patient

TABLE 4.1

Example Benefit-Risk Comparison of Three Hypothetical Treatments

Treatment Characteristics	Treatment A	Treatment B	Treatment C
Indication	Colorectal cancer	Epilepsy	Common cold
Number of patients per year	1,000	100,000	10 million
Annual risk	10 heart attacks per year (1%)	10 heart attacks per year (0.01%)	10 heart attacks per year (0.0001%)
Benefit	15-year increase in expected survival	Improvement from significant disability to near normal	Reduces cold duration by 2 days

subgroups. Because of the potential magnitude of the health and financial consequences of decisions based on these evaluations, enormous research resources are devoted to improving measurement of treatment-related benefits and harms. However, even if all the difficulties related to measuring benefits and harms could be resolved, there still would be no solution to the fundamental problem of how to compare dissimilar outcomes when making drug development and regulatory decisions, however precisely measured.[*]

While the quality of clinical evidence is important, the more fundamental question for drug development and regulatory decision making is how much risk of adverse outcomes, including mortality, is acceptable for a treatment that offers improvements in efficacy relative to the current standard of care. Consider, for example, the three hypothetical treatments shown in Table 4.1. Treatment A increases expected survival of each of 1,000 colorectal cancer patients by 15 years, treatment B significantly improves the quality of life of 100,000 epilepsy patients, and treatment C reduces the duration of the common cold by 2 days for 10 million patients. All three treatments will result in 10 heart attacks per year in the treated population.

Many people might agree that the benefit justifies the risk for treatment A, fewer might agree that treatment B is acceptable, and even fewer might agree that treatment C is acceptable. However, the reasons for these judgments are unrelated to the quality of the evidence presented. Rather, reasonable people might reasonably disagree about how much importance to attach to the outcomes and the size of the treated population. For some decision makers, a relatively small benefit in a sufficiently large population could offset a given adverse event risk, while for others, any risk of a fatal side effect would make a treatment unacceptable. These differences clearly involve subjective judgments. This chapter suggests that valid measures of patients' B-R trade-off

---

[*] In this chapter, we do not attempt to distinguish between B-R assessment for drug development decision making and B-R assessment for regulatory decision making. The likelihood that a product will obtain regulatory approval is a primary concern in pharmaceutical, biologic, and medical device development decisions.

preferences could provide systematic evidence to assist decision makers in evaluating such treatments.

4.1.2 Whose Preferences Should Count?

Balancing benefits and risks involves both technical assessments of the evidence base and societal value judgments about relative importance. Such evaluations typically are conducted by advisory bodies comprised of scientists and clinicians. It therefore is important to distinguish between evaluations for which there often are established scientific standards and subjective value judgments about which both experts and nonexperts reasonably could disagree.[1] Because patients are the potential beneficiaries of more effective treatments and also bear the risks associated with those treatments, their subjective judgments about relative importance arguably warrant serious consideration. However, current evaluation practices do not require quantification or even formal consideration of the values of patients in the treatment review and approval process.[*] The values and risk tolerance of patients are typically presented qualitatively to advisory panels and policy makers either individually or through advocacy organizations.

Patients sometimes have had a significant effect on drug development and regulatory decisions. For example, people with HIV successfully demanded early access to experimental antiretroviral treatments two decades ago.[3] Public testimony at U.S. Food and Drug Administration (FDA) advisory committee meetings sometimes includes forceful personal demands for expanded, informed access to treatments, as in the case of withdrawn products for irritable bowel syndrome and multiple sclerosis,[4,5] or for stronger warnings and restricted use, as in the case of antidepressants thought to increase the potential for suicidal ideation and behavior in some pediatric patients.[6] However, such anecdotal testimony does not provide systematic evidence of the willingness of well-informed patients to accept observed or expected risks to achieve the therapeutic benefits of these products. Furthermore, it is unclear whether those who advocate for less restrictive or more restrictive access to treatments are representative of the population for whom treatment is indicated.

Devising a more formal role for patient preferences in regulatory decision making may be hampered by the perception that patients are unwilling or unable to accept responsibility for their own treatment decisions. On the other hand, both the FDA and the European Medicines Agency (EMA) are developing institutional arrangements to provide more involvement for individual patients and patient organizations in regulatory decision making.[1,7,8] There is also growing interest in shared clinical decision making and accepting a more proactive role for patients. For example, in one study people

---

[*] An exception may be the regulatory guidance issued in 2012 by the FDA's Center for Devices and Radiological Health and Center for Biologics Evaluation and Research.[2]

with rheumatoid arthritis were asked who should decide which inhibitor of tumor necrosis factor (TNF)-alpha they should use.[9] Patients were evenly split between those wanting their rheumatologist to make the decision and those preferring either a joint decision between themselves and their rheumatologist or making the decision themselves alone.

The principle of informed consent clearly requires that patients play a meaningful role in therapeutic decisions. The American Medical Association's official statement on informed consent reads, in part:

> Informed consent is more than simply getting a patient to sign a written consent form. It is a process of communication between a patient and physician that results in the patient's authorization or agreement to undergo a specific medical intervention. In the communications process, you ... should disclose and discuss [specific details on potential benefits and risks of treatment]. In turn, your patient should have an opportunity to ask questions to elicit a better understanding of the treatment or procedure, so that he or she can make an informed decision to proceed or to refuse a particular course of medical intervention.[10]

Informed consent and shared decision making shift the focus of decision making from physicians determining what is best for patients to physicians helping patients decide what is best for them. This shift of focus requires a deep understanding of what information is useful in aiding patient-centered decision making, how much information is enough, and what methods of communicating this information are most effective.

Despite formal acknowledgment of the importance of shared decision making, the patient perspective is unlikely to play a meaningful role in drug development and regulatory B-R evaluations in the absence of valid, quantitative evidence on patients' tolerance for treatment-related risks. Methods are needed for quantifying the value that patients place on the potential benefits of a treatment and their willingness to accept risk in return for those potential treatment benefits. An ideal solution would have both scientific credibility and be compatible with widely accepted approaches for evaluating evidence. Our objective is to present a pragmatic, but also rigorous method for systematically quantifying the risk tolerance of patients in a form likely to be useful for healthcare decision making. Understanding what kinds of side effect risks are of greatest concern to patients may help identify appropriate strategies for modifying drug formulations, designing more useful labels, helping physicians communicate more effectively with their patients, and helping inform more consistent and principled licensing decisions.

4.1.3 Benefit-Risk Thresholds

Regulatory authorities are developing improved patient outreach mechanisms to elicit qualitative reactions to proposed regulatory decisions.[8]

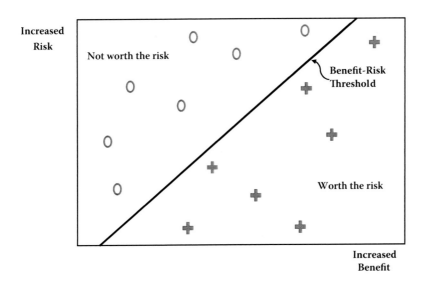

FIGURE 4.1
Linear benefit-risk threshold.

However, such mechanisms do not provide quantitative evidence to facilitate formal comparisons of therapeutic benefits with the risks of adverse
treatment events. For example, a tabulation of the number of positive and
negative comments from patients and patient organizations poses problems
of how to integrate and interpret the data. Statisticians might ask whether
the comments are representative of a particular population of patients.
Should comments be weighted by the intensity of preference expressed?
Should comments be weighted by how informed the individual is about the
available evidence? Should comments from patients who have experienced
adverse outcomes be weighted differently than patients who have experienced beneficial outcomes?

Comprehensive B-R analysis requires quantitative patient preference data
that are collected under experimentally controlled conditions. Obtaining
valid and reliable quantitative measures of patients' risk tolerance for
given therapeutic benefits requires first defining the constructs of interest.
Figure 4.1 illustrates the problem of evaluating trade-offs between benefits
and risks.* An index of therapeutic benefits derived from clinical data is
shown on the horizontal axis, and the observed or likely incidence of an
adverse therapeutic outcome (probability of a given adverse event) is shown
on the vertical axis. The B-R threshold can be interpreted as the rate at which
patients are willing to accept an increase in the adverse event risk in return
for an increase in treatment efficacy.

---

* The framework in Figure 4.1 is similar to that originally presented in the CIOMS IV report.[11]

People are averse to bearing risk unless they are sufficiently compensated with better efficacy, which requires that the threshold slope upward. Combinations of benefits and risks that lie above the line indicate outcomes for which the subjective value of the incremental risks (relative to a comparator) is greater than the corresponding subjective value of the incremental benefits. These points are indicated by O. The net benefits of these combinations (incremental benefits minus incremental risks) are negative, and thus would make individuals worse off relative to the comparator. Conversely, outcomes that lie below the line indicate outcomes for which the subjective value of the incremental risks is less than the corresponding subjective value of the incremental benefits (indicated by +). The net benefits of these combinations are positive, and thus would make individuals better off relative to the comparator. Holding efficacy constant at a given level, the threshold indicates the subjective risk value at which benefits are exactly offset by risks for that level of efficacy.

The threshold is likely to be nonlinear, following the standard economic law of diminishing marginal utility. A principle of economic theory widely confirmed in empirical studies is that the additional value of one more unit of a good decreases as the consumption of that good increases. (Think of the incremental satisfaction you get from the first cup of coffee of the day compared to the fourth cup.) Thus, we generally expect the B-R threshold to increase at a decreasing rate with increases in treatment benefit.

B-R thresholds are likely to differ across risk types. Figure 4.2 is an example of B-R thresholds estimated in a renal cell carcinoma (RCC) preference study.[12] For a given level of efficacy defined as increased number of months of progression-free survival, patients are willing to accept higher treatment-related risk of lung damage than liver damage. B-R thresholds also can differ among patients. Some patients may be more tolerant than others of side effect risks such as liver or lung damage as a result of inherent differences in their aversion to bearing risk, severity of their health condition, or other factors.

4.1.4 Using the Benefit-Risk Threshold to Compare Benefits and Risks

Many discussions about B-R comparisons focus on the need to convert benefits and harms into a common metric to facilitate direct calculation of net benefits. Finding such a metric is difficult and requires the quantitative evaluation of relative importance across benefits and harms. In contrast, the B-R trade-off threshold offers a framework for comparing quantitative measures of benefits and risks in several ways. Suppose Figure 4.3 is the B-R threshold for a specific patient population in a given therapeutic area. Suppose also that the benefits and risks associated with a treatment in that therapeutic area correspond to point A, with efficacy E_A and risk level R_A.[*]

---

[*] See Lynd and O'Brien[13] for an example of how such a point can be plotted using clinical trial data.

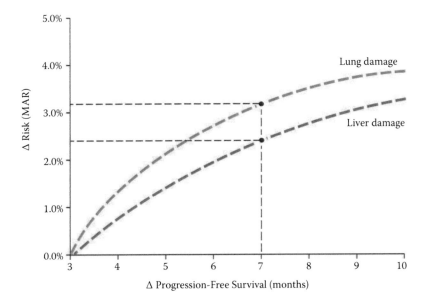

FIGURE 4.2
Benefit-risk thresholds for renal cell carcinoma.

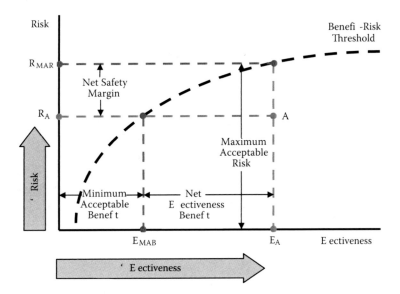

FIGURE 4.3
Quantitative measures of benefit-risk trade-offs.

Let R_{MAR} correspond to the maximum acceptable risk (MAR) patients in a specified population are willing to accept in exchange for an observed or expected treatment benefit represented by E_A. The difference between R_{MAR} and the observed or expected risk R_A is the net safety benefit, that is, the risk in excess of the observed or expected risk that these patients are willing to accept in exchange for that level of treatment efficacy.

Alternatively, acceptable trade-offs can be described in relation to efficacy. For the observed or expected level of risk (R_A), there is a corresponding level of efficacy on the trade-off curve (i.e., a minimum acceptable benefit (MAB) required for patients to accept this level of risk). MAB is measured as the horizontal distance between the vertical axis and the B-R threshold at R_A and is indicated by E_{MAB} on the horizontal axis. Because E_A is to the right of the B-R threshold, this treatment yields a net efficacy benefit equal to the difference between observed or expected efficacy E_A and MAB (E_{MAB}). If, for example, efficacy for oncology treatments was measured as progression-free survival, then the difference between actual and minimum acceptable progression-free survival is the net efficacy benefit measured as the average number of extra months provided by the therapy beyond the minimum number of months necessary to offset the effect of bearing a specific treatment risk.

The B-R trade-off curve also can be used in the context of more familiar measures used in clinical settings. Despite well-known limitations, number needed to harm (NNH) and number needed to treat (NNT) are used as measures of clinical outcomes.[14] For any given level of efficacy, MAR can be interpreted as the ratio of the maximum acceptable number of adverse events (N_{AE}) to the size of the exposed population (N). NNH is a measure of the number of people that need to be exposed to the treatment to yield one case of a specific adverse event assuming a constant therapeutic benefit. MAR is a preference-based measure elicited from a relevant group of patients, while NNH is a clinical measure of safety. Larger NNH values indicate greater benefit.[14] Thus, the inverse of MAR (1/MAR) is a measure of the minimum acceptable number needed to harm that can be compared directly with the clinical measure of NNH.

Conversely, MAB represents the minimum level of efficacy that patients require to accept a given level of risk. Thus, for a given level of adverse event risk, MAB is the ratio of the minimum number of patients achieving a therapeutic benefit (N_B) to the total size of the treated population (N). Likewise, NNT is a measure of the number of people that need to be exposed to the treatment to yield one case of therapeutic benefit, assuming a constant level of treatment-related risk. Smaller NNT values indicate greater benefit. Thus, the inverse of MAB (1/MAB) is a preference-based measure of the maximum acceptable number needed to treat that can be compared directly with the clinical measure NNT.

Comparing 1/MAR and 1/MAB to NNH and NNT, respectively, provides an indication of the extent to which patients find the B-R balance of a given treatment acceptable. Specifically, if NNH > 1/MAR, then patients, on average, are willing to accept the risk associated with a given level of therapeutic benefit. If NNT < 1/MAB, then patients, on average, consider the therapeutic benefit of the treatment to outweigh the treatment-related risk for a given level of risk. For example, Powell and Gibson[15] used NNT to indicate increased efficacy in a higher dose of inhaled corticosteroids for oral candidiasis side effect in treating patients with asthma. The NNH was 21 for a daily dose of 500 µg compared to an NNH of 90 for a daily dose of 100 µg.[15] If MAR is 2.5%, then there is a positive net benefit at 100 µg but a negative net benefit at 500 µg (1/0.025 = 40 and 90 > 40 > 21). Post-myocardial infarction prophylactic interventions, such as adding aspirin to streptokinase to reduce 5-week vascular mortality rates, may have NNTs as high as 20 to 40.[14] To yield positive net benefits, MAB would have to be less than 5% if NNT is 20 and less than 2.5% if NNT is 40.

4.1.5 Criteria for Valid Assessments of Patient Preferences

As discussed, the B-R trade-off threshold indicates the rate at which patients are willing to accept increases in risk for gains in benefits and allows for direct comparisons between risk tolerance and observed (or predicted) efficacy and adverse event incidences. Identifying this threshold for patients requires a set of subjective preference weights for changes in beneficial and harmful outcomes. Methods for obtaining these weights must satisfy several criteria for relevance and validity. These are listed in a seven-point validity checklist:

1. Respect patients' inherent attitudes and values. The principles of informed consent and shared decision making imply that patients' concerns should play a central role in healthcare decisions. While other stakeholders may have different perspectives, quantification of patient preferences should measure the preferences of sufficiently well-informed patients.

2. Employ widely accepted principles of effective risk communication. Effective risk communication is challenging given the low level of numeracy in the general population. In eliciting patient preferences, it is important to define the context of the B-R trade-offs, explain the severity and likelihood of treatment-related harms, and help patients conceptualize probabilities using appropriate numeric, verbal, and graphic representations of risk.

3. Minimize possible distortions resulting from framing and decision heuristics. The cognitive effects of framing (such as describing changes as gains or losses), anchoring (such as implicit signaling of a reference value), and simplifying heuristics (such as recoding numerical values to low, medium, and high) have been well documented.[16] Best-practice survey techniques minimize such effects by using strategies that include:

- Mimic realistic clinical decision frames as closely as possible. For example, treatment decisions have to be made before actual efficacy and side effect outcomes are realized. Presenting such outcomes as likelihoods relative to no treatment or current treatment replicates that decision frame.
- Avoid suggesting "reasonable" ranges of outcomes in the background information and randomize the question sequence.
- Include "cheap talk" text. Cheap talk is text that breaches the usual anonymity between survey researchers and respondents to engage respondents in the research problem and to motivate them to devote more effort to the preference elicitation task than they otherwise would.[17]

4. Incorporate measured endpoints from clinical data. Preferences must be measured over relevant clinical domains to be useful in evaluating available evidence. However, clinical endpoints often take the form of surrogate markers (e.g., liver enzymes) that may be asymptomatic. In such cases, patients must be helped to understand how such measures affect the likelihood of more serious outcomes.

5. Identify the relative importance of clinical endpoints for a sufficiently well-informed and sufficiently representative sample of patients. Acceptable evidence standards require that preferences be measured for representative samples of sufficiently large numbers of patients to support generalization to the population of interest. If subgroup analysis is required, the sample must include sufficient numbers in each subgroup to test for differences between groups of patients. Also, the sample's preferences should be sufficiently well informed to ensure correspondence between preference weights and measured clinical endpoints.

6. Ensure that preference weights satisfy basic utility-theoretic requirements of logic and consistency. The data should include internal validity tests and should be verified for conformity with logic, consistency, and other utility-theoretic requirements.

7. Account for nonlinearities in trade-off preferences. Although nonlinearity is not a required property of well-defined preferences, linearity should not be assumed. Study designs should allow for expression of nonlinear preferences and threshold effects. Continuous, linear relations should be used only after appropriate specification tests.

4.1.6 Quantifying Patient Preferences Using Stated Choice Methods

Well-designed stated choice surveys, also known as discrete choice experiments or choice-format conjoint analysis, offer an excellent available approach for eliciting preference weights required to construct B-R trade-off curves. Stated choice methods recognize that goods and services have value because of their characteristics or attributes. People's relative preferences for these attributes vary, and they usually are willing to accept trade-offs among them. Stated choice surveys ask respondents to indicate their preferences for various attribute combinations through a series of judgments under controlled experimental conditions. Appropriate statistical analysis of trade-off data reveals the weights people assign to various treatment attributes. Thus, an important advantage of stated choice methods is that they can provide values for individual benefit and side effect endpoints as well as for healthcare interventions as a whole. Stated choice methods also facilitate internal checks for attentiveness, consistency, and logic because each respondent provides answers to multiple stated choice questions. Such studies also can satisfy all the other requirements in the preceding list of preference elicitation requirements for relevance and validity.

Analysts use stated choice methods to quantify preferences for a variety of market and nonmarket goods and services. These include medical interventions, pharmaceutical treatments, and environmental health risks, as well as marketed goods and services.[18-25] Although the literature on health applications is relatively small when compared to the number of applications in other areas (e.g., transportation, environmental economics), health researchers are able to draw upon over 30 years of well-established good-practice research standards.

Implementing a stated choice study that satisfies the seven validity requirements noted above requires successful execution of several steps to develop and field the survey, analyze the survey data, and apply the estimated preference weights to calculate metrics from comparing benefits and risks. The remainder of this chapter considers each of these steps in the context of a detailed example for metastatic renal cell carcinoma.

4.2 Quantifying Patient Preferences for Benefit-Risk Trade-Offs: An Empirical Example

We will use an empirical example to illustrate the design and implementation of a stated choice survey to quantify patients' B-R trade-off preferences using a study of patient preferences for metastatic renal cell carcinoma (RCC) treatments.[12]

4.2.1 Evaluating the RCC Benefit-Risk Study Using the Seven-Point Validity Checklist

The RCC study satisfies the seven criteria for valid assessments of patient preferences.

First, during survey development, the researchers invited 15 patients with a diagnosis of RCC to participate in face-to-face interviews to ensure that patients understood the definitions of attributes and levels, accepted the hypothetical context of the survey, and understood the survey language. The survey instrument was revised based on findings from the pretest interviews before it was finalized and administered online to a larger group of patients with RCC.

Second, a risk grid was used to illustrate the chance of having a serious adverse event when taking an RCC medication. After a brief tutorial, nearly all patients (96%) answered the test question correctly. A risk grid interpretation question was included in the survey after a risk tutorial to test respondents' understanding of the information conveyed in the risk grid graphics.

Third, to minimize possible distortion from framing or decision heuristics, the survey instrument was carefully pretested, and a best-practice experimental design technique was used to balance task complexity and statistical efficiency given the intended sample size. The survey took about 25–30 minutes to complete on average, and each respondent answered 10 choice questions. The respondent burden was considered acceptable according to the studies that report the relationship between respondent burden and response error variance.[26]

Fourth, the RCC study incorporated clinical endpoints in selecting attributes and determining the levels of the attributes. The results of the preference study were then linked to trial data to predict choice probabilities between two comparators that were used to treat patients with RCC in clinical trials.

Fifth, the study sample included a sufficiently large and diverse sample to test for variations in patients' risk tolerance. The inclusion criteria were adults living in the United States with a self-reported physician diagnosis of RCC. The respondents were recruited from a patient support group, the Kidney Cancer Association. The researchers recruited patients at all stages of RCC, as people with nonmetastatic RCC may become future metastatic RCC patients. Responses from a total of 272 respondents were included for data analysis. Forty-three percent of the sampled patients had metastatic RCC. The statistical analysis found no differences in preferences between patients with metastatic and nonmetastatic RCC.

Sixth, the estimated preference weights satisfy basic internal validity requirements of logic and consistency. For example, better clinical outcomes logically should be preferred to worse clinical outcomes. Results from the RCC study indicated that 10 months of progression-free survival (PFS) had a higher preference weight than 5 months of PFS, and 5 months of PFS had a higher preference weight than 3 months of PFS. Also, mild-to-moderate side effects were preferred to severe side effects.

Finally, the preferences indicated nonlinearity consistent with diminishing marginal utility. For example, for the three levels of PFS included in the study (3 months, 5 months, and 10 months), the slope for an additional month

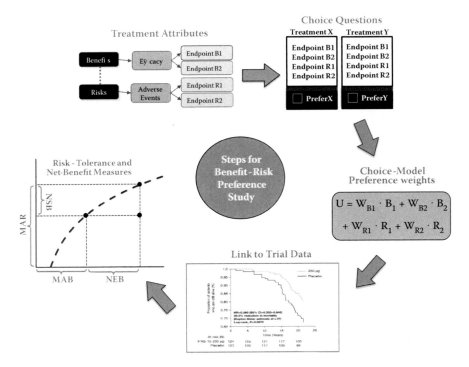

FIGURE 4.4
Required steps for a stated choice study.

of PFS differs in the two segments. An additional month of PFS in the segment between 3 months and 5 months has a marginal utility of 0.89, while an additional month of PFS in the segment between 5 months and 10 months has a marginal utility of 0.47.* Thus, an additional month of PFS was more valuable to patients with a baseline PFS range of 3 to 4 months.

4.2.2 Steps Required to Implement a Stated Choice Study

Figure 4.4 illustrates the various steps required to implement a stated choice study of patients' B-R preferences.

Implementing a valid and reliable B-R preference study involves several considerations, including identifying the set of clinical trial endpoints (attributes and attribute levels) patients will evaluate, constructing a series of choice-format B-R trade-off questions, estimating an appropriate statistical choice model to obtain relative preference weights, applying the resulting weights to measured endpoint data, and obtaining risk tolerance and net benefit estimates. We briefly discuss each of these steps below. Readers

---

* We use the term *utility* here with the conventional economic meaning of neoclassical ordinal utility rather than cardinal utility scaled between zero and one that is used commonly in estimating health state utilities.

interested in learning more about stated choice methods can find extensive details in Hensher et al.[27]

4.2.2.1 Identify a Subset of Clinical Trial Endpoints

The list of study treatment factors or attributes must be both clinically relevant and meaningful to patients. Trial data often contain too many endpoints for patients to evaluate simultaneously. However, the list of potential attributes can be reduced to a tractable size by considering factors that are likely to influence drug development, clinical decisions, or regulatory outcomes. In B-R preference assessments designed to provide evidence for regulatory decision making, efficacy endpoints typically include primary and secondary efficacy endpoints measured in clinical trials. Adverse event attributes should include those adverse events that likely will be of concern to regulators. Patient focus groups or interviews are necessary to confirm the relevance of the study attributes to patients.

The attributes for the RCC study shown in Table 4.2 included PFS, which was the primary endpoint in the clinical trials, as well as two potential serious adverse events: lung damage (pneumonitis) and liver failure (hepatic impairment). Furthermore, in designing the study, the researchers matched the descriptions of the levels of the tolerability side effects (fatigue, stomach problems, mucositis/stomatitis, and hand-foot syndrome) to the Common Terminology Criteria for Adverse Events (CTCAE) version 4.02 applied in clinical trials (National Cancer Institute).[28] For example, "mild to moderate" corresponded to grades 1 and 2 and "severe" corresponded to grade 3.

4.2.2.2 Construct a Series of Trade-Off Questions

The most common choice-question format uses two or three hypothetical treatment alternatives. An example choice question from the RCC study is shown in Figure 4.5, including risk grids to minimize numeracy requirements for comparing probabilistic outcomes. In a stated choice survey, each treatment profile specifies benefit and risk endpoints corresponding to ranges observed in the clinical data. Combinations of endpoints are varied according to an experimental design with known statistical properties. Each respondent evaluates several questions to obtain enough data to ensure feasible statistical estimation of all the desired preference weights. Marshall et al.[26] found that most published stated choice studies have used between 7 and 15 questions per respondent. The resulting survey instrument must be carefully pretested with patients from the target sample population to ensure that the information provided in preparation for the choice questions is adequate and understandable, that patients are willing to accept trade-offs among the ranges of treatment attributes shown, and that the overall burden of taking the survey is acceptable.

TABLE 4.2

Treatment Attributes and Levels for Metastatic Renal Cell Carcinoma

Attribute	Levels
How long the treatment will keep the cancer from getting worse (progression-free survival)	10 months 5 months 3 months
Feeling weak or tired (fatigue)	None Mild to moderate Severe
Stomach problems (GI side effects)	None Mild to moderate Severe
Sores in your mouth or throat (mucositis or stomatitis)	None Mild to moderate Severe
Redness or sores on your hands and feet (hand-foot syndrome)	None Mild to moderate Severe
Chance of lung damage (pneumonitis)	No chance 5 out of 1,000 (0.5%) 10 out of 1,000 (1.0%) 20 out of 1,000 (2.0%)
Chance of liver failure (hepatic impairment)	No chance 5 out of 1,000 (0.5%) 10 out of 1,000 (1.0%) 20 out of 1,000 (2.0%)
How you take the treatment (mode of administration)	1 pill once a day with or without food 2 pills twice a day without food Infusion once a week

4.2.2.3 Estimate the Preference Weights

Statistical analysis of the pattern of choices observed from appropriately designed study reveals the implicit preference weights employed by respondents in evaluating the hypothetical treatments. Choice questions generate cross section/time series data that require analysis using advanced statistical techniques.[29,30] The pattern of choices observed from a sufficiently well-informed sample of patients contains information on the implicit weights respondents used to evaluate the relative importance of the outcomes. These weights are estimated using an appropriate statistical model that accounts for the particular characteristics of the choice data. In the RCC study, increasing PFS from 5 months to 10 months is worth more than twice as much as reducing liver failure risk from 0.5% to nil.

Subgroup analysis and other statistical techniques help identify how risk tolerance varies by individual characteristics such as age, health history, current symptom severity, treatment experience, and time since diagnosis. In the RCC study, male patients strongly preferred taking daily oral medications to

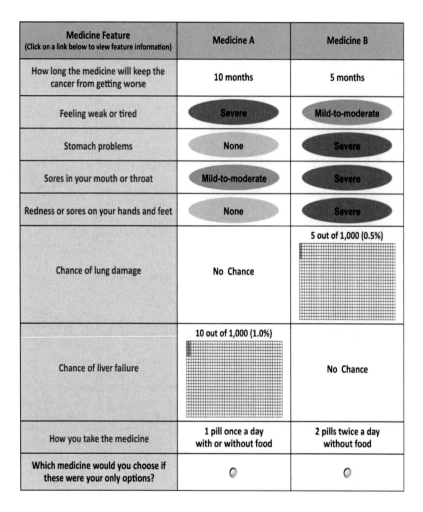

Medicine Feature (Click on a link below to view feature information)	Medicine A	Medicine B
How long the medicine will keep the cancer from getting worse	10 months	5 months
Feeling weak or tired	Severe	Mild-to-moderate
Stomach problems	None	Severe
Sores in your mouth or throat	Mild-to-moderate	Severe
Redness or sores on your hands and feet	None	Severe
Chance of lung damage	No Chance	5 out of 1,000 (0.5%)
Chance of liver failure	10 out of 1,000 (1.0%)	No Chance
How you take the medicine	1 pill once a day with or without food	2 pills twice a day without food
Which medicine would you choose if these were your only options?	○	○

FIGURE 4.5
Example of benefit-risk trade-off question.

weekly infusions, while female patients were indifferent between administration modes. Unfortunately, it is rarely possible to obtain a strictly random sample from the full population of patients with a particular condition. Furthermore, the sample must be sufficiently large and diverse to enable statistical weighting of estimates to approximate those of a more representative sample or identify preferences for groups of patients with particular characteristics of clinical interest.

4.2.2.4 Link the Preference Weights to Measured Endpoint Data

To evaluate the preference-weighted benefits and adverse event outcomes of one or more drugs, the preference weights must be combined with results from clinical or observational studies. It is important to note that combining preference weights with measured endpoint data is limited to the attributes measured in the stated choice survey. Because no relative preference weights exist for outcomes that are not included in the stated choice survey, nothing can be said about the relative importance of nonincluded outcomes. This requires the implicit assumption that outcomes not included in the stated choice survey are not relevant to the decision that the B-R assessment is designed to inform.

The RCC study evaluated outcome profiles for two drugs. Using the preference weights estimated in the stated choice model, weighted by observed outcome rates such as mean efficacy and incidences of side effects at different levels of severity, predicted percentages of patients who would prefer each drug were calculated. Table 4.3 compares the distributions of clinical trial outcomes for two main vascular endothelial growth factor (VEGF) tyrosine kinase inhibitors (TKIs), sorafenib and pazopanib (both approved second-line treatments for patients failing on a first-line cytokine-based treatment such as sunitnib), as well as their predicted choice probabilities. The choice probabilities in the table indicate the predicted proportion of patients in our sample who would prefer the expected outcomes of each treatment. These estimates are based on the specific attributes and levels that patients evaluated in the study.[12] The calculations therefore require an implicit assumption that all other features are the same between the two treatments.

Based on the information in Table 4.3, pazopanib was predicted to be chosen by two-thirds of the RCC patients in the sample, while sorafenib was predicted to be chosen by one-third of the patients. Pazopanib was the preferred treatment primarily because of its longer median PFS (an additional 1.9 months). The order of the preference scores could be reversed for patients who were more averse to liver-failure risk, which is lower for sorafenib.

Preference weights are measured on an ordinal scale, so individual weights cannot be compared directly; an outcome with preference weight of 0.8 is not twice as important as an outcome with preference weight of 0.4. However, we can compare relative *differences* in weights. Thus, we can say an improvement indicated by a difference in preference weights of 0.8 is twice as important as an improvement indicated by a difference in preference weights of 0.4. Figure 4.6 illustrates an MAR calculation for an improvement in PFS from 5 months to 10 months. The increase in efficacy results in an improvement of 0.84 in the preference weight score. Zero chance of liver failure has a score of 1.4. Subtracting 0.84 gives a score of about 0.56, a bit less than the loss resulting from an increase in liver failure risk from zero to 2%.

Using this approach, one can calculate the maximum acceptable risk patients were willing to accept on average for an additional month of PFS. Assuming a baseline PFS of 3–4 months (corresponding to placebo), patients would be willing to accept up to a 1.0% (95% CI: 0.8–1.4%) risk of lung damage and 0.7% (95% CI: 0.4–1.0%) risk of liver failure for an additional month

TABLE 4.3

Clinical Outcomes and Predicted Choice Probabilities for Metastatic Renal Cell
Carcinoma Drug Profiles

Attribute	Sorafenib (Nexavar)	Pazopanib (Votrient)
How long the medicine will keep the cancer from getting worse (progression-free survival)	5.5 months	7.4 months
Feeling weak or tired (fatigue)		
None	63%	81%
Mild to moderate	32%	17%
Severe	5%	2%
Stomach problems		
None	57%	48%
Mild to moderate	41%	49%
Severe	2%	3%
Sores in your mouth or throat (mucositis or stomatitis)		
None	100%	100%
Mild to moderate	0%	0%
Severe	0%	0%
Redness or sores on your hands and feet (hand-foot syndrome)		
None	70%	94%
Mild to moderate	24%	6%
Severe	6%	0%
Chance of lung damage (pneumonitis)	No chance	No chance
Chance of liver failure (hepatic impairment)	No chance	0.2%
How you take the medicine (mode of administration)	2 pills twice a day without food	1 pill once a day with or without food
Predicted choice probability	**31%**	**69%**

Source: Nexavar. FDA prescribing information. 2007. http://www.accessdata.fda.gov/drug-satfda_docs/label/2007/021923S004S005S006S007lbl.pdf. Votrient. FDA prescribing information. 2009. http://www.accessdata.fda.gov/drugsatfda_docs/label/2009/022465lbl.pdf.

of PFS. Thus, if a new drug offers an additional month of PFS with an additional risk of liver failure of 0.5%, then the incremental benefits outweigh the incremental risks. There is a positive net benefit because patients, on average, would be willing to accept *up to* an additional 0.7% risk of liver failure for the additional month of PFS. The four B-R trade-off measures shown in Figure 4.3 calculated with the estimated preference weights for PFS and liver damage are as follows:

- The maximum acceptable risk of liver failure for the additional 1.9 months of PFS is 0.4%.
- The net safety benefit is 0.2%.

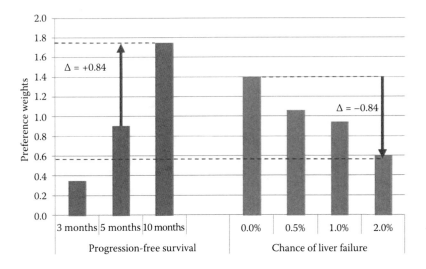

FIGURE 4.6
Example MAR calculation in the RCC study.

- The minimum acceptable benefit for the additional 0.2% risk of hepatic impairment is 24 days of PFS.
- The net efficacy benefit is an additional 33 days of PFS.

4.3 Discussion

In this chapter, we have described a method of quantifying patients' B-R trade-off preferences. While the stated choice approach described here is a promising method for developing valid and reliable evidence of patients' views, as with all methods, it is subject to some important limitations.

The most serious limitation of stated preference methods is that they employ judgments among hypothetical alternatives. Hypothetical choices do not involve the same clinical, financial, and emotional consequences as actual choices. Potential hypothetical bias can be limited by constructing choice questions that mimic realistic clinical choices as closely as possible and map clearly into clinical evidence.

Successful elicitation of patient preference data requires that all members of the patient sample share a common, clear understanding of the profile features they are evaluating. In the case of eliciting preferences, including probabilistic outcomes, researchers must cope with the generally low level of numeracy in the general population. Some well-established findings include Glickman and Gough[31] and Kahneman and Tversky:[32]

- People have difficulty evaluating small probabilities (smaller than about 1 in 1,000) accurately.
- People care as much about such risk characteristics as voluntariness, familiarity, catastrophe, dreadfulness, and timing as they do about probability.
- People evaluate risk information from different sources differently.
- People evaluate decisions framed as gains differently than decisions framed as losses.
- People translate population risks to personal risks using a variety of heuristics.

Attitudes toward bearing risk and related behavior depend strongly on the decision context, sources of risk, and content of available information, as well as specific characteristics of the risks in question. When such attitudes and heuristics are likely to be used in actual healthcare decision making, they could be a legitimate expression of preferences.

While the literature of stated choice applications in healthcare generally is significant, there are only 30 to 40 published studies specifically related to B-R trade-off preferences. Many clinicians, regulators, and other potential users of B-R assessments therefore are unlikely to be familiar with these methods and may be skeptical of their validity and reliability. While obtaining valid and reliable measures of patient preferences is challenging, it is possible to quantify patient preferences and subject the results to the same rigorous standards as those applied to clinical, epidemiological, and patient-reported outcomes data.

B-R evaluations inevitably require assessments of both data quality and the relative importance of endpoints measured in clinical data. Relative importance requires value judgments about which both experts and nonexperts could reasonably disagree. Incorporating quantitative patient perspectives on these value judgments is consistent with the growing interest in greater patient involvement in healthcare decision making. While public outreach efforts may give voice to patient concerns about regulatory decisions, it is not obvious how such outreach efforts can and should influence regulatory policy and decision making. To be relevant and useful, preferences should be quantified and evaluated in the same way as other forms of evidence in evaluating new and existing therapies. Eliciting and quantifying the preferences of patients will allow for more formal, evidence-based consideration of patient perspectives that is currently lacking in regulatory decision making.

References

1. European Medicines Agency. 2013. *Guidance document on the content of the Rapporteur day 80 critical assessment report.* http://www.ema.europa.eu/docs/en_GB/document_library/Regulatory_and_procedural_guideline/2009/10/WC500004800.pdf.
2. Center for Devices and Radiological Health. 2012. *Factors to consider when making benefit-risk determinations in medical device premarket approval and de novo classifications.* http://www.fda.gov/downloads/MedicalDevices/DeviceRegulationandGuidance/GuidanceDocuments/UCM296379.pdf.
3. Arno, P., and Feiden, K. 1992. *Against the odds: The story of drug development, politics and profits.* New York: Harper Collins.
4. Armstrong, D. 2005. Research bolsters tarnished MS drug. *Wall Street Journal,* April 13, 2005.
5. Mathews, A., and Westphal, S. 2006. Tricky FDA debate: Should a risky drug be approved again? *Wall Street Journal,* February 24, 2006.
6. Grady, D., and Gardiner, H. 2004. Over prescribing prompted warning on antidepressants. *New York Times,* March 24, 2004.
7. European Medicines Agency. 2010. Road map to 2015. http://www.ema.europa.eu/docs/en_GB/document_library/Report/2011/01/WC500101373.pdf.
8. Maxmen, A. 2012. Law spurs regulator to heed patients' priorities. *Nature* 487: 154.
9. Chilton, F., and Collett, R. 2008. Treatment choices, preferences and decision-making by patients with rheumatoid arthritis. *Musculoskeletal Care* 6: 1–14.
10. American Medical Association. 1998. Informed consent. http://www.ama-assn.org/ama/pub/physician-resources/legal-topics/patient-physician-relationship-topics/informed-consent.page.
11. CIOMS IV Working Group. 1998. *Benefit-risk balance for marketed drugs: Evaluating safety signals.* Report of the CIOMS Working Group IV. Council for International Organizations of Medical Sciences, Geneva.
12. Wong, M., et al. 2012. Patients rank toxicity against progression free survival in second-line treatment of advanced renal cell carcinoma. *Journal of Medical Economics* 15: 1139–1148.
13. Lynd, L., and O'Brien, B. 2004. Advances in risk-benefit evaluation using probabilistic simulation methods: An application to the prophylaxis of deep vein thrombosis. *Journal of Clinical Epidemiology* 57: 795–803.
14. McQuay, H., and Moore, R. 1997. Using numerical results from systematic reviews in clinical practice. *Annals of Internal Medicine* 126: 712–720.
15. Powell, H., and Gibson, P. 2003. Inhaled corticosteroid doses in asthma: An evidence-based approach. *Medical Journal of Australia* 178: 223–225.
16. Tversky, A., and Kahneman, D. 1981. The framing of decisions and the psychology of choice. *Science* 211: 453–458.
17. Ozdemir, S., Johnson, F., and Hauber, B. 2009. Hypothetical bias, cheap talk, and stated willingness to pay for health care. *Journal of Health Economics* 28: 894–901.
18. Wittink, D., and Cattin, P. 1989. Commercial use of conjoint analysis: An update. *Journal of Marketing* 53: 91–96.

19. Viscusi, W., Magat, W., and Huber, J. 1991. Pricing environmental health risks: Survey assessments of risk-risk and risk-dollar trade-offs for chronic bronchitis. *Journal of Environmental Economics and Management* 21(1): 32–51.
20. Johnson, F., and Desvousges, W. 1997. Estimating stated preferences with rated-pair data: Environmental, health, and employment effects of energy programs. *Journal of Environmental Economics and Management* 34: 79–99.
21. Johnson, F., Fries, E., and Banzhaf, H. 1997. Valuing morbidity: An integration of the willingness-to-pay and health-status index literatures. *Journal of Health Economics* 16: 641–665.
22. Ryan, M., and Hughes, J. 1997. Using conjoint analysis to assess women's preferences for miscarriage management. *Health Economics* 6: 261–273.
23. Bryan, S., Buxton, M., Sheldon, R., and Grant, A. 1998. Magnetic resonance imaging for the investigation of knee injuries: An investigation of preferences. *Health Economics* 7: 595–603.
24. Johnson, F., Desvousges, W., Ruby, M., Stieb, D., and DeCivita, P. 1998. Eliciting stated health preferences: An application to willingness to pay for longevity. *Medical Decision Making* 18: S57–S67.
25. Johnson, F., Banzhaf, M., and Desvousges, W. 2000. Willingness to pay for improved respiratory and cardiovascular health: A multiple-format, stated-preference approach. *Health Economics* 9: 295–317.
26. Marshall, D., et al. 2010. Conjoint analysis applications in health—How are studies being designed and reported? An update on current practice in the published literature between 2005 and 2008. *Patient: Patient-Centered Outcomes Research* 3: 249–256.
27. Hensher, D., Rose, J., and Greene, W. 2005. *Applied choice analysis.* New York: Cambridge University Press.
28. National Cancer Institute. 2010. Common terminology criteria for adverse events, v. 4.0. http://ctep.cancer.gov/protocolDevelopment/electronic_applications/ctc.htm.
29. Train, K. 2003. *Discrete choice models with simulation.* New York: Cambridge University Press.
30. Train, K., and Sonnier, G. 2005. Mixed logit with bounded distributions of correlated partworths. *Applications of Simulation Methods in Environmental and Resource Economics* 6: 117–134.
31. Glickman, T., and Gough, M. 1990. *Readings in risk.* Washington, DC: Resources for the Future Press.
32. Kahneman, D., and Tversky, A. 2000. *Choices, values, and frames.* New York: Cambridge University Press.

5

Benefit-Risk Modeling of Medicinal Products: Methods and Applications

Lawrence D. Phillips

CONTENTS

5.1 Introduction

Concern about insufficient transparency, communicability, and consistency of drug decision making by regulators led four senior European drug regulators[1] to write in the *New England Journal of Medicine*:

> Regulators need to refine their methods of assessing benefit-risk balances and switch from "implicit" to "explicit" decision making—that is, to an approach involving explicit descriptions not only of all deci-

sion criteria and interpretations of data but also valuations, such as the weighting factors for potential treatment outcomes.

Ideally, regulators should also shift from the use of qualitative statements to quantitative descriptions of the size of the net health benefits.

Exploration of explicit quantitative approaches began to appear two decades earlier: Ciba-Geigy sponsored a 2-day workshop at the London School of Economics in 1989, followed by another workshop in 1990 sponsored by the Centre for Medicines Research (CMR), both of which brought together drug regulators with managers from the pharmaceutical industry to explore how quantitative modeling might assist drug decision makers.[2] Although quantitative models developed on the spot in both seminars demonstrated the potential for combining different types and sources of data with clinical judgment, it was clear that improving and making explicit decisions about drugs were not widely considered to be serious concerns.

Several years later, the Council for International Organizations of Medical Sciences (CIOMS IV, 1999)[3] argued for a systematic approach to benefit-risk (B-R) assessment "to strengthen decision making in the interest of public health." By the middle of the first decade of this century the topic began to rise in importance, so CMR sponsored two further workshops in 2004 and 2005, which eventually led to the publication of "a new approach" for assessing the B-R balance of drugs,[4] followed by those authors' book, which provided a more detailed exposition.[5]

A clear need to give drug decision makers improved processes for assessing the B-R balance of drugs emerged from a meeting in November 2007 of regulators, industry professionals, and academics. The Next Steps Working Group (a smaller group now numbering about 25 members) formed to explore how best to carry forward the work begun at the 2007 meeting. Largely as a result of their efforts, a B-R Assessment Working Group was established in April 2012 as part of the Drug Information Association's Clinical Safety and Pharmacovigilance Special Interest Area Community (DIA SIAC).

In 2009, the European Medicines Agency (EMA) established a 3-year research program whose purpose was "to develop and test tools and processes for balancing multiple benefits and risks as an aid to informed regulatory decisions about medicinal products." Then, in September 2009, a public-private partnership of 33 organizations, coordinated by the EMA, began the 5-year Pharmacoepidemiological Research on Outcomes of Therapeutics by a European Consortium (PROTECT) program, sponsored by the Innovative Medicine Initiative (IMI) and the European Federation of Pharmaceutical Industries and Associations (EFPIA). The goal of the project was to strengthen the monitoring of the B-R of medicines throughout their life cycle.

However, the move from implicit to explicit, and from qualitative to quantitative assessments, has not met with universal acclaim. Five objections have been made against quantitative models at workshops and conference presentations, with proponents offering the following responses to each:

"The B-R balance for drug X was obvious from looking at the data." True for some people, but not for all, especially those who focus on one or more unfavorable effects. The research by Andrea Beyer during the EMA's Benefit-Risk Project established substantial differences in risk attitude among 80 European assessors.[6] For example, safety assessors were found to be consistently more risk averse than efficacy assessors across many different products. This research also found gender differences in risk attitude—female assessors generally saw more benefit than male assessors for 28 types of medicinal products. In reviewing a mock dossier, assessors were found to be influenced by their worry about safety, the number of people exposed to the risk, and ethical issues associated with the drug. An important conclusion was that the probability of an adverse event and its severity are not the only factors considered by assessors. Thus, what seems to be obvious about a drug for one person may not be true for another.

"I don't need a quantitative model; it can all be done based on experience and judgment." The research to date makes clear that both experience and judgment are required, but they are subjective and not necessarily easily expressed in words, making communication difficult about why the B-R balance was judged as reported. This makes it difficult to train young assessors, especially as the judgments of older assessors will inevitably be based on different experiences. In part, these different experience bases may explain why the U.S. Food and Drug Administration (FDA) and EMA do not always agree about a new drug. Decision analysts find that words can hide disagreements that surface when numbers are assessed.[7] Indeed, the numbers often facilitate finding the right words to explain the differences, which can then be discussed, resolved, and explained to inexperienced assessors.

Another aspect of this objection is that it confuses confidence in one's decisions with accuracy. Decisions about drugs are predictions about future effectiveness in actual use. Clinical trials are fallible indicators of effectiveness, and the high confidence of decision makers is not necessarily an accurate indicator of that effectiveness. Confidence in the B-R balance can lead to an "illusion of validity."[8]

"It's too subjective." This criticism is usually levied at models that distinguish between objective data and their subjective interpretation. Clinical trials provide objective data, and many models summarize the data in ways that aid decisions, leaving the decision maker to judge the clinical relevance of the data for the purpose at hand. But the data describe objective qualities of a drug, such as quality, efficacy, and safety; understanding benefits and risks requires interpretation of the data's relevance in the context of treatment. In addition, the B-R balance depends on subjective judgments to establish what

effects to consider, how they will be evaluated, how relatively important they are, who is perceiving the importance, and other factors that are unique for every intended decision and its anticipated outcome. Thus, the B-R balance depends on both objective data and their subjective interpretation. The "too subjective" criticism assumes that the B-R balance can be determined by wholly objective means. But the most comprehensive models make explicit many of those subjective factors in a way that subjects them to scrutiny and facilitates communication about why the B-R balance was judged to be favorable or unfavorable.

"You can't represent the benefit-risk balance with a single number." This comment presumes there is no common unit for making comparisons across different effects, but there is: preference value (or utility), long known in decision theory[9] and extensively used in multicriteria decision analysis (MCDA).[10] Utilities also appear in quality-adjusted life years (QALYs),[11] as well as a variety of multicriteria models that are limited to health outcomes and may or may not include risk criteria. That said, the question also assumes there is just one B-R balance, whereas there are as many as there are assumptions and judgments associated with a model. The point of modeling is to see how different scenarios affect the results, which collectively provide a sound basis for the choices of decision makers.

"Clinical judgment is too complex to be captured in numbers." This criticism is at the heart of a 60-year-old debate provoked by the clinical psychologist Paul Meehl in his "disturbing little book," *Clinical versus Statistical Prediction*, first published in 1954. He reported evidence that simple, linear, additive models, or algorithms, consistently outperformed clinical predictions of behavior. This finding, based on the 22 studies available at the time, dropped a bombshell in psychology that still reverberates today. By 1996, the original sample of 22 studies had grown to 136 comparative studies, just 8 of which favored clinical prediction.[12] This led Grove and Meehl[13] to conclude that mechanical combination of judgments "is almost invariably equal to or superior to the clinical method." Daniel Kahneman[8] brought this line of research up-to-date (p. 223):

The number of studies reporting comparisons of clinical and statistical predictions has increased to roughly two hundred, but the score in the contest between algorithms and humans has not changed. About 60% of the studies have shown significantly better accuracy for the algorithms. The other comparisons scored a draw in accuracy, but a tie is tantamount to a win for the statistical rules, which are normally much less expensive to use than expert judgment. No exception has been convincingly documented.

In summary, consideration of these five objections reveals perspectives that do not fit easily with the objectives of B-R modeling. Nevertheless, resistance to explicit, quantitative approaches persists. Several working hypotheses about this state of affairs will be discussed later in this chapter.

First, however, it will help to explore some limitations of the current B-R decision-making process, see how the EMA has clarified the meaning of benefits and risks, and examine the capability of decision theory models to capture both data and clinical judgments in modeling the B-R balance of drugs. A case study of a new drug that was modeled during its approval process will illustrate how a quantitative model works. The chapter concludes by introducing a potentially serious bias in judging or modeling the B-R balance.

5.2 How Regulators Assess New Medicinal Products

The starting point for any research project on decision making within a particular domain is to find out how decisions are currently made. To see how this is done in Europe, the EMA's Benefit-Risk Project team interviewed over 50 regulators and experts in six European National Competent Authorities (NCAs), a drug regulatory agency. Three findings stand out:[14] First, a new drug application, whose size can be more or less 10 gigabytes, is distributed by an NCA to teams or individuals responsible for producing the overall B-R recommendation. Experts are engaged to assess various parts of the application, and the parts are brought together by the responsible individual or group. Second, the B-R balance is assessed intuitively, as a result of extensive consultation and discussion, and a final decision by an NCA is made by consensus or by voting. No quantitative model is employed by any of the agencies (nor, as it turns out, by any agency anywhere in the world). Third, the meanings of *benefit* and *risk* are very fluid. Each interviewee was asked, "What does your agency/you think of as a benefit?" and the same question was posed for a risk. The benefit question elicited 37 different words or phrases, while 51 different words or phrases were given to the risk question. The former were generally aligned to favorable effects, with some exceptions, but many of the answers to the risk question were incompatible (e.g., "tolerance of a drug compared to serious side effects," "severity of side effects," "frequency of side effects").

Since good science requires clear definitions of key variables, the Benefit-Risk Project team recommended that the language in the final B-R section of an assessment report be clarified to distinguish the magnitude of effects from the uncertainties of those effects, represented in the fourfold table shown in Figure 5.1. These distinctions were adopted in September

Favorable effects	Uncertainty of favorable effects
Unfavorable effects	Uncertainty of unfavorable effects

FIGURE 5.1
The fourfold model of benefits and risks of drugs.

2010,[15] accompanied by clear guidance for each of the four cells. This chapter will use the model to avoid possible ambiguity about which cell is being addressed.

5.3 Tools and Methods for Benefit-Risk Assessment

Since CIOMS IV, many surveys of the literature on B-R approaches have appeared. Most have focused mainly on describing the various methods, with little, if any, attempt to assess their suitability for B-R assessment.[3,4,16,17] The EMA survey,[18] which included all approaches in previous surveys, covered 3 qualitative frameworks and 18 quantitative methods. That survey assessed the methods against five major objectives: logical soundness, comprehensiveness, acceptability of results, practicality, and generativeness (usefulness). Each of these objectives was clearly defined by three to five bullet points, for a total of 21 criteria. Under the headings "Our View," the key advantages and disadvantages for the approaches were explained, with particular emphasis on their relevance for regulatory decision making. The approaches were a mixture that defied easy classification: some were models, others were statistical methods, and a few were measurement techniques. A subsequent PROTECT methodology review[19] reports a classification system of four headings: frameworks (with subcategories of descriptive frameworks and quantitative frameworks), metrics (threshold indices, health indices, and trade-off indices), estimation techniques, and utility survey techniques. Altogether, that survey used 19 criteria to describe 47 approaches. Of these 47, the authors of the report recommended 13: 2 descriptive frameworks (BRAT (from the Benefit-Risk Action Team) and PrOACT-URL (Problem Formulation, Objectives, Alternatives, Consequences, Trade-Offs, Uncertainties, Risk Attitude, and Linked Decisions)), 2 quantitative frameworks (MCDA and stochastic multicriteria acceptability analysis (SMAA)), 6 metric indices, 2 estimation techniques, and 1 utility survey technique (conjoint analysis). These 13 were considered by the PROTECT teams that modeled six drugs. For the purposes of this chapter, BRAT, PrOACT-URL, MCDA, and SMAA were confirmed as the most comprehensive frameworks.

At this writing, 15 identified drugs have been modeled by applying one or more of the surveyed methods, as shown in Table 5.1.

TABLE 5.1

Named Drugs Subjected to Quantitative Modeling by Mid-2013

	Drug	**Indication**
London School of Economics	Rimonabant	Obesity
MSc students	Certolizumab pegol	Rheumatoid arthritis
Post-marketing models	Sunitinib	Gastrointestinal cancer
	Lapatinib	Breast cancer
EMA Benefit-Risk Project	Tafamidis	Transthyretin amyloidosis
Pre-marketing models	Briakinumab	Psoriasis
	Vandetanib	Medullary thyroid cancer
	Tocilizumab	Juvenile idiopathic arthritis
	Belimumab	Systemic lupus erythematosus
PROTECT	Efalizumab	Psoriasis
Post-marketing models	Natalizumab	Relapsing remitting multiple sclerosis
	Rimonabant	Obesity
	Rosiglitazone	Type II diabetes
	Teithromycin	Antibiotic
	Warfarin	Stroke prevention

The experience of modeling the four drugs under active consideration by the EMA made clear that the use of 17 or 21 criteria was a case of overkill.[20] The task of being explicit and quantitative about the B-R balance means that numerical representations are required for both data and its clinical relevance, for the trade-offs between benefits and risks, and for all uncertainties. The task itself provides just four pragmatic criteria:

1. **Comprehensive:** Provides for any number of favorable and unfavorable effects to be considered and accepts any performance measures (measurable quantities, scoring systems, relative frequencies, health outcomes, etc.).

2. **Separates fact from judgment:** Distinguishes measured data about favorable and unfavorable effects from the clinical relevance of the effects.

3. **Provides a common unit:** Quantifies clinical judgments about clinical relevance and about trade-offs between the effects, thereby establishing a common unit that enables comparison of combinations of effects, and can aggregate all effects.

4. **Accommodates uncertainty:** Quantifies uncertainty about all effects.

A fully comprehensive, explicit, and quantitative model would satisfy all four criteria, but most of the approaches considered by the EMA fell short (see Table 5.2). Judgments about this were based on reading the literature about all the approaches and on experience reported about the 15 B-R models

TABLE 5.2

Extent to Which Benefit-Risk Approaches Satisfy the Four Pragmatic Criteria in the Text

Fully Satisfies	Partially Satisfies	Fails to Satisfy
Decision trees with multiattribute valuations	Bayesian statistics (including Markov processes)	Principle of threes
MCDA with sensitivity analysis or probabilistic simulation	Probabilistic simulation	NNT/NNH
SMAA	Kaplan-Meier estimator	Relative risk ratios
	Q-TWiST	TURBO
	QALYs	Contingent valuation
	Incremental net health benefit	
	Stated choice methods	
	Stated preference methods	
	Evidence-based benefit and risk model	

listed in Table 5.1. Table 5.3 gives a short description of each approach and the key reasons for its classifications in Table 5.2.

Considering that benefits and risks are defined by the multiple criteria against which drugs are evaluated, it is not surprising that many of the approaches can be classed more generally under the heading of multicriteria analysis.[21,22] Although Belton and Stewart[21] use multicriteria decision analysis as a generic term for the many methods they explain in their book, in this chapter MCDA is used in the narrower sense of any multicriteria approach that is based on the axiomatic system leading to expected utility theory.

The application of any quantitative method is necessarily based on a qualitative framework that sets out the steps to be followed to arrive at a final balancing of benefits against risks. These concepts are introduced in Chapter 1 and one of them, PrOACT-URL, will be applied to an illustrative example later in this chapter.

5.4 Decision Theory and Social Process

The three approaches in the left column of Table 5.2 are all based on decision theory. Decision analysis, an intellectual technology based on decision theory, starts with one simple assumption: people often wish to make coherent choices. The principle of coherent choice was first given an axiomatic foundation by Frank Ramsey (1926)[23] and further developed by von Neumann and Morgenstern (1947)[24] and Savage (1954).[25] Raiffa and Schlaifer (1961)[26] provided the full integration between the probability calculus for modeling uncertainty and the utility theory for expressing the relative value of outcomes and the decision maker's risk attitude. Ron Howard provided the

TABLE 5.3

The Approaches in Table 5.2 Defined with Their Relevance for Benefit-Risk Assessment

Approach/Method	Description	Relevance for Benefit-Risk Balance
Decision trees and influence/ relevance diagrams	Models of choices, subsequent uncertain events, further decisions, etc., and consequences	Best for problems dominated by uncertainty. Based on decision theory, which ensures coherence of decisions. Most drug cases not dominated by a single effect, so benefit and risk consequences will be multiattributed.
Multicriteria decision analysis (MCDA)	Extends decision theory to accommodate multiple, conflicting objectives	Best for problems dominated by multiple effects. Provides a common unit of preference value for both benefits and risks. Incorporates evidence and judgments about the clinical relevance of the evidence. Uncertainty usually handled with sensitivity analysis.
Stochastic multicriteria acceptability analysis (SMAA)	MCDA with probabilistic simulation (the latter described below)	Fixed input values for MCDA can be replaced with probability distributions. Monte Carlo analysis gives probability distributions of B-R balance.
Bayesian statistics	Statistical inference with probabilities representing degrees of belief	Integrates evidence and its uncertainty, both pre- and post-approval, with multiple effects in decision models, thereby providing updating of degrees of belief as evidence becomes available.
Markov processes	Conditional probabilities express movement over time from state to state	A useful addition to the above to capture processes such as the progression of a disease in a patient over time. Can model the long-term benefits and risks of treatment.
Probabilistic simulation	Uncertainties about effects represented with probability distributions	Separate probability distributions for favorable and unfavorable effects, with an overall probability distribution for the benefit-risk balance obtained with Monte Carlo simulation.
Kaplan-Meier estimator	Graph showing proportion of patients surviving over time for drug and comparator treatments	Relevant to displaying changes in health states over time. Can be used in Markov models or decision trees, and can incorporate the utilities of the health states.
Quality-adjusted time without symptoms or toxicity (Q-TWiST)	Extends QALYs and Kaplan-Meier graphs to model effect trade-offs	Shows balance between prolongation of life and suffering from treatment toxicity. Can accommodate utilities, but is not a full MCDA.

Continued

TABLE 5.3 (*Continued*)

The Approaches in Table 5.2 Defined with Their Relevance for Benefit-Risk Assessment

Approach/Method	Description	Relevance for Benefit-Risk Balance
Quality-adjusted life years (QALYs)	Multicriteria models of preferences or desirability of health outcomes	Focus on health outcomes ensures relevance to public health decisions about resource allocation, but is insufficient for drug development or approval, with its focus on efficacy and safety outcomes.
Incremental net health benefit (INHB)	An MCDA model of differences of favorable and unfavorable effects	Has mainly focused on health outcomes and their uncertainties, using relative value-adjusted life-years (RVALYs), but can use QALYs or utilities or any other health outcome metric.
Stated choice methods, e.g., conjoint analysis or discrete choice experiments	Measurement methods for determining preference weights from hypothetical choices	Focuses on trade-offs between effects, particularly relevant to eliciting patients' preferences. Preference weights are not the same as MCDA swing weights. Doesn't consider uncertainty.
Stated preference methods	Various methods for assessing patients' utility functions	Rating scale, standard gamble, and time trade-off techniques for assessing utilities and probabilities can be relevant to any of the above models.
Evidence-based benefit and risk model	Data for each effect shown as 3 generic dimensions of a box	A simple multicriteria model that is prescriptive about the dimensions and is restricted in scope. B-R balance given by sums of weights.
Principle of threes	Effects assessed using 4-point scales	Too restricted and simplistic for even moderately complex B-R assessment, and sum of scores fails to recognize relative importance of effects.
Number needed to treat (NNT) Number needed to harm (NNH)	The inverse of the difference between treatment and comparator in proportions of an effect (favorable or unfavorable)	These statistics don't take into account the clinical relevance of the effects, provide no means for balancing benefits against risks, and their focus on probability differences can lead to logically unsound decisions.
Relative risk ratios	Ratios of effect probabilities for treatment/comparator	Ratios cannot be combined meaningfully to inform the B-R balance. No account taken of the clinical relevance of effects.
Transparent uniform risk-benefit overview (TURBO)	Two favorable and two unfavorable effects scored and weighted	A simple multicriteria model with four effects is too restricted for most drug decisions.
Contingent valuation	Effects translated into monetary values as the common metric	Based on economists' cost-benefit analysis. Monetary values obtained through willingness to pay or to sell studies. Of questionable relevance.

name *decision analysis* along with a systems approach for implementation,[27] and Howard Raiffa provided the detail about how to assess probabilities and utilities and conduct a decision analysis.[9] Almost a decade later, Keeney and Raiffa (1976)[10] extended decision theory to accommodate multiple objectives with their associated criteria. More than 20 years after their incorporation of multiple objectives, Keeney and Raiffa joined with John Hammond to write an accessible layperson's view of how to apply decision analysis within their new eight-step PrOACT-URL framework.[28]

In practice, decision analysis helps decision makers construct coherent preferences and provides a foundation for an intelligent and rational conversation among key players who can contribute their differing perspectives. Sometimes it is used to improve the quality of decisions, but it is more frequently employed to help decision makers and key players arrive at a shared understanding of relevant issues, develop a sense of common purpose, and agree on the way forward.[29] This latter purpose is often referred to as sociotechnical decision analysis and is a blend of technical decision-analytic modeling with social processes that ensures that the right people are making the right contributions at the right time. Key stakeholders explore how results change with different assumptions, contrasting judgments, imprecision of data, and differences of opinion, thereby providing support to decision makers as they make the final decision. Ideally, this is carried out in face-to-face workshops,[30] guided by an impartial and experienced facilitator who is a specialist in on-the-spot modeling.

This chapter provides a social framework that enables the key players to work together, using MCDA as a shared language for expressing preference values and uncertainty (Figure 5.1). The framework illustrates how the B-R balance can be decomposed into its constituent pieces, and how these pieces may be reassembled into an overall balance. The process can be expedited with the help of specialized software specifically designed for MCDA modeling, so no mathematical expertise is required of participants. Commercially available software that implements MCDA consistent with the Keeney-Raiffa approach includes Hiview,[*,31] V.I.S.A.,[32] Logical Decisions,[33] and M-Macbeth.[34]

5.5 An Agenda for Multicriteria Decision Analysis

MCDA is a methodology for appraising alternatives on individual, often conflicting, criteria and combining them into one overall appraisal. Table 5.4 presents the steps in creating and exploring an MCDA model. These steps were inspired by Figure 6.1 of *Multi-Criteria Analysis: A Manual* (Dodgson

---

[*] Hiview3 was used to model drug X and display graphically all the results reported here.

TABLE 5.4

Steps in Creating and Exploring an MCDA Model

Step	Task
Context 1. Establish the decision context	• Identify the medicinal product (e.g., new or marketed chemical or biological entity, device, generic). • Identify the therapeutic area and indication for use. • Recognize the unmet medical need, severity and morbidity of condition, affected population, patients' and physicians' concerns, time frame for health outcomes. • Define the decision problem (what is to be decided and by whom).
Alternatives 2. Identify the options.	• Describe the medicinal product (e.g., new drug, drug at different doses, drug plus adjunct, device). • Describe the comparators (e.g., placebo, competitive products, gold standard).
Criteria 3. Identify and define the criteria for assessing the effects of each alternative. Represent these in an effects tree (referred to as attribute tree in Chapter 1).	• Select the favorable effects (e.g., endpoints, relevant health states, clinical outcomes, survival). • Select the unfavorable effects (e.g., adverse events, serious adverse events, infections, serious infections, major cardiac events, strokes, bone fractures, flares, etc.).
Weighting 4. Establish a measurement scale for each criterion and assess the relative importance of the scales.	• Define each effect's measurement scale and its units (e.g., mean or median scores, proportions, incidence, change from baseline, frequency of events per patient) and determine upper and lower limits that encompass a plausible range for the data. • Assess swing weights to represent the clinical relevance of the swing from the lower to the upper limit on each scale.
Scoring 5. Describe how the alternatives perform for each of the criteria and show how to convert input data into preference values (i.e., assess value functions).	• Gather available data, pooling or performing meta-analyses of multiple data sources, to give data summaries and confidence intervals. • Provide data summaries in an effects table with alternatives in columns and criteria in rows. • Assess linear or nonlinear value functions, usually direct (more means better) for favorable effects, and inverse (more means worse) for unfavorable effects.
Results 6. Calculate results and provide graphical displays.	• Multiply preference values and criterion weights and sum the products to obtain overall value (usually carried out by appropriate software). • Construct preference value bar graphs for favorable and unfavorable effects, and for individual effects. • Calculate difference displays for pairs of alternatives.

TABLE 5.4 (*Continued*)

Steps in Creating and Exploring an MCDA Model

Step	Task
Sensitivity analyses 7. Explore effects of uncertainty on the benefit-risk balance.	• Vary individual weights over their entire range from 0 to 1.0; display the overall results graphically. • Change input data over ranges of uncertainty (e.g., pessimistic values for favorable effects and optimistic ones for unfavorable effects). • Examine the overall B-R balance under possible future scenarios (e.g., adverse events, shifts in unmet medical need) by changing input data and criteria weights. • Revise any of the above numbered steps and tasks as insights emerge.
Recommendation 8. Formulate recommendations.	• Judge the relative importance and effect of the decision maker's risk tolerance for this product (e.g., orphan drug status, special population, unmet medical need, risk management plan). • Consider how this decision is consistent with similar past decisions, might set a precedent, or make similar decisions in the future easier or more difficult. • Metabolize the results before making any decisions (newly constructed preferences can change with reflection as new insights surface).

et al., 2000)[22] and have been supplemented by elements of the PrOACT-URL framework and by practical experience to apply specifically to modeling the B-R balance of a drug and its comparator(s).

5.6 An Illustrative Example

An example will clarify how the MCDA framework of Table 5.4 can be applied.[*] The following is for a hypothetical drug X, which is based on an amalgam of real drugs. Although a variety of experts and clinicians, working in small groups or consulted individually, carried out most of the tasks listed in Table 5.4, this narrative draws upon experience facilitating workshops for groups in pharmaceutical companies and regulatory agencies exploring or applying quantitative modeling for assessing the B-R balance of a drug.[20,35]

---

[*] This is a more detailed version of the drug X case study reported in EMA (2011),[20] but here with slightly different weights and, initially, a concave value function for malignancies and a convex one later.

A meeting was conducted as a 1-day facilitated workshop,[30] called a decision conference.[29] The drug X model was constructed on the spot during the decision conference, with the model projected onto a screen so everyone could see each step in its construction. Participants included pharmaceutical professionals and regulators, and represented a variety of perspectives and experience.

5.6.1 Context

The decision conference started by exploring the issues concerning drug X, a pharmaceutical treatment for moderate to severe, active rheumatoid arthritis in adult patients. Drug X was intended for use with methotrexate (MTX) when the response to antirheumatic drugs (including MTX) was inadequate. The medical need for treatment for more difficult cases of rheumatoid arthritis is well established. The group agreed to role-play regulators faced with assessing the B-R balance so they could make appropriate recommendations about drug X's approval.

5.6.2 Alternatives

The group considered three pharmaceutical alternatives, each delivered by injection, two of which represented drug X at two different doses:

1. Placebo—MTX only
2. Drug X—200 mg plus MTX
3. Drug X—400 mg plus MTX

5.6.3 Criteria and Effects Tree

The group spent some time constructing the effects tree. To start, they were unclear in reading relevant European Public Assessment Reports (EPARs) about which of many effects were taken into account in arriving at a final judgment about the B-R balance. The publicly available information reported numerous favorable effects (FEs) and unfavorable effects (UFEs), but did not make clear which of these should inform the judgment about the B-R balance of drug X (although ACR20* was given as the primary endpoint, with secondary endpoints of ACR50 and ACR70). The facilitator suggested that the group consider only those effects that discriminate the alternatives from a clinical point of view. It was difficult to decide which unfavorable effects should be

---

* ACR20, ACR50, and ACR70 refer to classifications of the American College of Rheumatology by which patients are classified as either responders, if they have shown a certain improvement in their rheumatoid arthritis, relative to a baseline assessment, or nonresponders, if they have not. The values 20, 50, and 70 indicate progressively more significant levels of improvement.

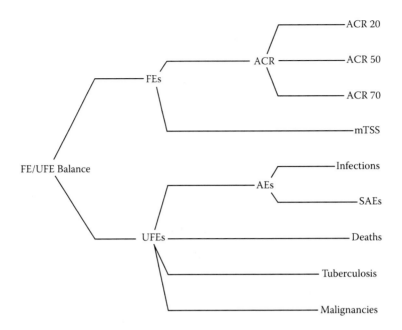

FIGURE 5.2
The final effects tree for the drug X benefit-risk assessment, represented here by the FE/UFE balance, with favorable effects shown under the FEs node and unfavorable effects under the UFEs node.

taken into account, as a considerable number of side effects were reported. The clinicians in the group contributed their views about which they would consider to be of importance, a task that others found more difficult. The original effects tree included more effects than the final tree shown in Figure 5.2; the tree trimming occurred at various stages of weighting and scoring.

The definition of the ACR score took some time to pin down as it wasn't included in the EPAR, but it was eventually discovered in a Web search, which established that it is a multicriteria scoring system.

Ideally, criteria in an MCDA should meet the following requirements (see Section 5.4.4 of Dodgson et al. (2000)[22] for more detail):

1. **Completeness:** All criteria that could affect the overall result are included.

2. **Nonredundant:** Unnecessary or duplicate criteria have been removed.

3. **Operational:** Each alternative's effects can be judged or measured for each criterion on a scale that is monotonic in preference (criteria are defined so that either larger or smaller effects are preferred to intermediate effects).

4. **Preference independent:** The criteria are mutually preference independent. That is, the preference order of options on any one criterion

is unaffected by the preference order on any other criterion. This condition is weaker than statistical independence. Scores on criteria can be statistically correlated but preference independent. This is a requirement if weighted scores are to be interpreted unambiguously, not just for MCDA.

5. **Double counting:** For a given effect, a patient's response should be counted only once (this is frequently violated in B-R assessment of drugs; see below).

6. **Requisite size:** Only as many criteria are included in the analysis as is necessary to arrive at a decision.

7. **Time:** Time can be included as appropriate in the definitions of criteria; different time horizons can be established for different criteria.

MCDA is quite robust to violations of these requirements, but it is wise where possible to test that robustness in sensitivity analyses. It is at this stage, structuring the B-R problem, that expertise in defining the criteria and building the effects tree, along with clinical experience, can make the most important contributions.

For drug X all these criteria (and more) had been considered in the EPAR, so the group took them as given. However, one participant expressed concern that a patient who was counted in the ACR70 effect also appeared in the other two, less stringent ACR effects, thereby violating the double-counting criterion. If this were allowed, then the addition of ACR30 and ACR50 effects would lead that same person to be counted three times, thereby enhancing the benefit side of the equation. Ideally, the three ACR categories should have been defined as ranges, ACR20–39, ACR40–69, ACR70 and above, a simple frequency distribution. However, the facilitator felt that for now, the effects actually considered by the regulators should be honored, and that reducing the effects of double counting could be simulated by reducing the weight on the ACR node. As will be seen later, that assumption turned out to be incorrect.

After creating the effects tree, the facilitator directed the group to identify a metric for each effect (e.g., percentage, change score, number) along with upper and lower limits of each fixed scale. These limits created a plausible range for the data. Limits for similar metrics typically differ from one effect to another, much like an indoor thermometer and an oven thermometer share the same metric but use different ranges of temperatures. Two principles for establishing the ranges are in slight conflict. First, each range should be sufficiently narrow to ensure realism, which will facilitate the swing-weighting process (unrealistic values make it difficult to judge the clinical relevance of a swing in preference value when one or both limits defining the range have never been achieved by any other drugs). Second, it is more difficult to assess swing weights when the ranges for the same metric are unique for each effect. A compromise between similar ranges for related effects and unique ones for wholly separate effects is usually required.

In summary, creating the effects tree, defining the effects, identifying metrics for the effects, and establishing plausible data ranges for the scales must all be completed before the process of weighting the criteria can begin.

Upon completing a report of a decision conference, all relevant information, judgment, and data are summarized in an effects table. However, to help the reader follow the steps in its creation, a completed effects table is provided as a touch point as Table 5.5.

5.6.4 Weighting

The weighting process began by considering the three ACR effects, shown in Figure 5.3. The task confronting the group was to judge the swing from 0 to 100% on each of the scales, by answering the question: "How big is the difference, and how much do you care about it?" (without regard to the data associated with each of the alternatives). The difference is the same for all three effects, 100 percentage points, but caring about the difference is a matter of clinical relevance. When asked whether they would prefer a drug whose added clinical value extended to 100% of patients achieving an ACR score of 20, 50, or 70, the group quickly agreed that everyone achieving a 70 was best, so a weight of 100 was assigned to the ACR70 effect scale. Achieving 100% on ACR50 was also quite good, so participants agreed to assign a weight of 70. At first, many gave low weights to the ACR20 scale because that level of achievement is barely important clinically, but one participant argued for a higher weight on the grounds that this scale was the primary endpoint, so the group agreed on a weight of 40.

Next, the group considered weights at the next higher level, the FE node. The group compared the ACR effect scale with the highest weight, ACR70, with the mTSS* scale (Figure 5.4). The group judged the mTSS swing from a change score of 10 to 0 to be equal in clinical relevance to the 0–100 percentage point swing on the ACR70 scale. How is it possible to compare a change score with a percentage? By considering the clinical added value: an mTSS change score improvement from 10 to 0 is as clinically desirable as an improvement from 0 to 100% of patients achieving an ACR of 70. It is the *clinical relevance* of one preference value change compared to the other, not the changes themselves, that is compared.

Weights are scale constants, showing the trade-off between units on one scale compared to another. Here, a swing in preference value of 10 units for mTSS (from 10 to 0) equates to a swing in preference value of 100 percentage points for ACR70 (from 0 to 100%). These weights are associated with the ranges of the scales only, not necessarily with increments within those ranges.

The group then turned to the UFEs, first weighting the two adverse event (AE) effect scales. The serious adverse event (SAE) swing was judged to be

---

TABLE 5.5

Effects Table for Drug X, Giving Definitions of the Effects, Ranges of the Measurement Scales, Swing Weights for the Ranges, and the Data

	Name	Description	Units	Fixed Lower[a]	Fixed Upper[a]	Scale Weight	Placebo	Drug X 200 mg + MTX	Drug X 400 mg + MTX
Favorable effects	ACR20 (primary endpoint)	Proportion of patients achieving ACR20 at week 24	%	0	100	7.4	11.7	58.2	59.6
	ACR50 (secondary endpoint)	Proportion of patients achieving ACR50 at week 24	%	0	100	12.9	5.8	34.8	36.6
	ACR70 (secondary endpoint)	Proportion of patients achieving ACR70 at week 24	%	0	100	18.5	2.4	18.8	16.1
	mTSS (co-primary endpoint)	Mean amount of progression of joint damage in hands and feet at week 52	Mean change Score ± SD	0	10	18.5	2.8 ± 7.8	0.4 ± 5.7	0.0 ± 4.8

Unfavorable effects								
Infections	Proportion of patients experiencing infections and infestations	Number per 100 patient-years	70	80	3.9	72.13	79.88	76.62
SAEs	Proportion of patients experiencing musculoskeletal and connective tissue disorders	Number per 100 patient-years	25	60	5.5	57.05	28.39	25.88
Deaths	Proportion of patient deaths	%	0	3	18.5	0.15	0.42	0.97
Tuberculosis	Number of patients contracting tuberculosis	Number	0	30	5.5	0	5	28
Malignancies	Proportion of patients developing at least one malignancy	%	0	2	9.2	0.9	1.9	1.4

[a] Lower to upper limits define the range of a measurement scale that includes all the data for each criterion and is meaningful for assessing swing weights.

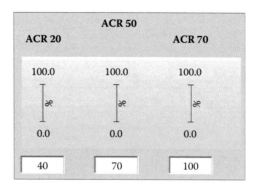

FIGURE 5.3
Relative swing weights assessed for the ACR scales.

the most clinically desirable, so it was given a weight of 100, and infections a weight of 70. Then the swing for SAEs was compared to the remaining three UFEs. The swing on deaths, from 3% to 0%, was judged to be the largest swing in preference, so it was given a weight of 100. Compared to the swing on deaths, the SAE swing was given a weight of 30, consequently reducing the infections weight to 30% of 70 and the SAEs weight to 30% of 100. Finally, compared to 100 for deaths, the group judged the tuberculosis swing to be 30 and the malignancies swing to be 50.

At this stage, all the FEs had been weighted relative to one another, and all the UFEs had been weighted relative to one another. All that remained was to compare the swings on the FE and UFE 100-weighted scales (ACR70 for FEs and deaths for UFEs). A swing from 0 to 100% of patients achieving ACR70 versus a reduction in deaths from 3% to 0% is shown in Figure 5.5. Which of these is most clinically desirable? The group agreed both swings were very important, and it was a tough call. After several minutes of debate,

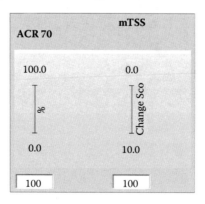

FIGURE 5.4
Relative swing weights assessed for the mTSS scale compared to the ACR70 scale.

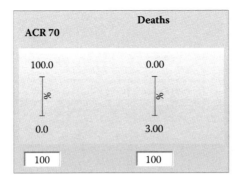

FIGURE 5.5
Swing weights comparing the 0 to 100% improvement for ACR70 to a reduction in deaths from 3% to 0%.

participants felt the swings were about equal, so weights of 100 were assigned to both.

With all the swing weights assessed, the weights at each node were multiplied along each branch of the effects tree to give terminal node products. The products were then summed, and each product was divided by the sum to yield the cumulative weights shown (multiplied by 100) in the "Scale Weight" column of the effects table. These show the added clinical value of swinging from the lower to the upper metric on each criterion, and therefore illustrate the relative importance of each range rather than the effect itself.

Note that the weighting process starts at the rightmost nodes, then progresses from right to left on the effects tree, always identifying the criterion of those compared to find the one associated with the largest swing in preference, and then judging the swings on the other criteria with that 100-weighted swing. Figure 5.6 shows the effects tree with the assessed swing weights, the terminal node products, and the normalized scale weights.

5.6.5 Scoring and Value Functions

The first task in scoring the options was to gather the data. Work in advance of the decision conference identified two clinical studies in which drug X was combined with MTX compared to MTX alone. In both studies, the ACR20 response at week 52 was taken as a primary endpoint, with secondary endpoints of ACR50, ACR70, and mTSS. Safety data came from four well-controlled, double-blind phase III studies in adult patients with rheumatoid arthritis. Questions naturally arose about how to pool the studies to arrive at a single statistical summary and 95% confidence interval for each statistic. Various approaches are available, but all use weighting systems that control the extent to which each study contributes to the overall summary;

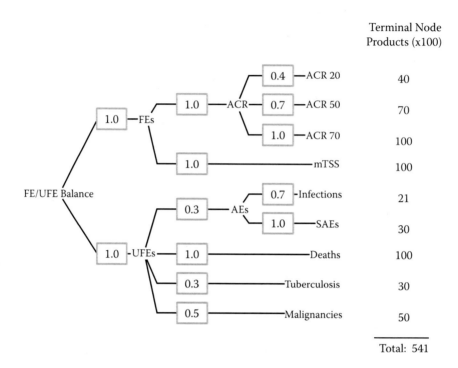

Terminal Node
Products (x100)

FIGURE 5.6
Effects tree with assessed swing weights, terminal node products, and normalized weights. Swing weights displayed just to the right of the FEs node compare the 100-weighted ACR criterion, ACR70, with the swing on the mTSS scale. Swing weights displayed just to the right of the UFEs node compare the 100-weighted AEs criterion, SAEs, with swings on deaths, tuberculosis, and malignancies.

specialized software can do most of the technical work.* Once those figures were obtained, the summaries were entered into the MCDA computer program and summarized in the effects table. When this was done for drug X, some of the data fell outside the ranges, so these ranges were widened to include all the data, accompanied by revisions of the weights.

The second task was to decide how those data were to be converted to preference values. The easiest method is a direct, linear conversion, as shown in Figure 5.7. The left figure shows the input data for the ACR20 effect displayed on the 0–100 percentage scale, while the right figure shows the conversion to 0–100 preference values. The scales are identical and the relative spacing of the three options remains unchanged. Figure 5.8 shows an inverse function for infections. Here, the scales are not the same numbers and the relative spacing has been inverted.

A linear conversion preserves relative differences in preference. The group felt that this was not appropriate for malignancies. Considerable debate

---

* For example, comprehensive meta-analysis from http://www.meta-analysis.com.

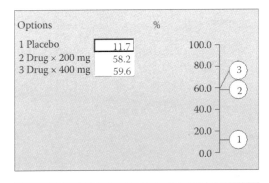

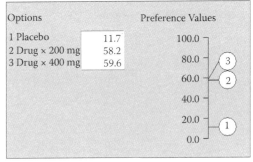

FIGURE 5.7
Direct, linear conversion for ACR20 of input percentage data (left) to 0–100 preference values (right).

followed, as a slight majority of participants argued for a concave function that tolerated a small proportion of malignancies, whereas the others thought the function should be convex and not tolerate even small proportions. After considering the graph of a linear function anchored at the limits of the range for malignancies (0 and 2%), the participants moved points on the graph upward to produce the concave function shown in Figure 5.9.

Ideally, no patients would experience malignancies and 2% or more would be so unacceptable that drug X would not be approved. If one-half of 1% did, then the preference value would be affected only slightly; a dip to 90 seemed appropriate. A further increase to 1% dropped the preference value a further 15 points to 75, while the next half-percent increase in malignancies reduced preference by 30 points to 45, with the largest drop from there to zero at the last half-percent increase to 2%.* Later, the group used sensitivity analysis to explore the effects of using a convex function.

---

* Value functions can also be created from qualitative judgments of differences in preference using the M-MACBETH function incorporated in Hiview3, or using MACBETH software at www.m-macbeth.com.

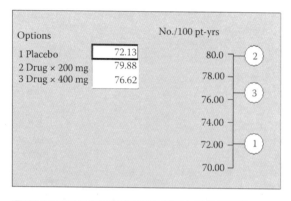

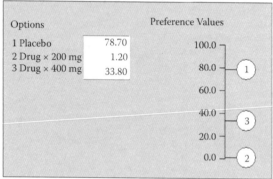

FIGURE 5.8
Indirect, linear conversion for infections of number per 100 patient-years to 0–100 preference values.

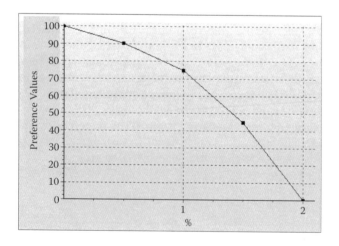

FIGURE 5.9
The concave function linking preference value to the fraction of patients contracting malignancies.

5.6.6 Results

The overall preference value for an option is a weighted average of the preference values on the nine effect criteria, as given in this equation:

$$\text{Total } pv_i = \sum_j w_j \times pv_{ij}.$$

The preference value pv_{ij} for option i on effect criterion j is multiplied by the cumulative weight on effect criterion w_j, then repeated for all j criteria. These products are then summed to give a total weighted preference value for the option. The total weighted preference value for each drug reflects the data about the drug's effects (the input scores) as well as the clinical relevance of the effects (the value functions and the criteria weights).

The weighted preference values for the three options are provided under the bar graphs in Figure 5.10. The 200 mg drug shows the greatest overall weighted preference value, followed by the 400 mg drug, with placebo least preferred. The lengths of the bars are proportional to the overall preference values, decomposed to depict the contributions of the favorable and unfavorable effects. As expected, placebo shows the least benefit (shortest green) and the greatest safety (longest red). The two doses of drug look about equal for favorable effects, but the 200 mg dose is slightly safer.

The separate contribution of each effect to the total weighted preference value is shown in Figure 5.11. Comparing identically colored slices enables a

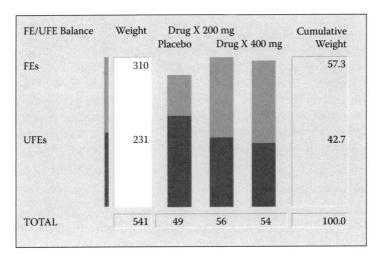

FE/UFE Balance	Weight	Drug X 200 mg			Cumulative
		Placebo	Drug X 400 mg		Weight
FEs	310				57.3
UFEs	231				42.7
TOTAL	541	49	56	54	100.0

FIGURE 5.10 (SEE COLOR INSERT)
Overall results are shown in the "Total" row. Longer upper (green) sections mean more benefit, while longer lower (red) sections indicate more safety. The two numbers in the "Weight" column are the sums of the nonnormalized input weights (terminal node products from Figure 5.6) for the FEs and UFEs, while the "Cumulative Weight" column shows the normalized weights.

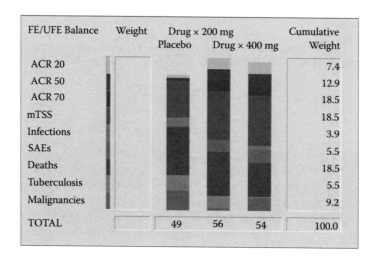

| FE/UFE Balance | Weight | Drug × 200 mg | | Cumulative |
		Placebo	Drug × 400 mg	Weight
ACR 20				7.4
ACR 50				12.9
ACR 70				18.5
mTSS				18.5
Infections				3.9
SAEs				5.5
Deaths				18.5
Tuberculosis				5.5
Malignancies				9.2
TOTAL		49	56 54	100.0

FIGURE 5.11 (SEE COLOR INSERT)
The contributions of each effect to the overall weighted preference values.

quick understanding of the main differences among the alternatives: SAEs are better for both doses compared to the placebo; the 400 mg dose is less safe with regard to tuberculosis; the 200 mg dose is less safe for malignancies.

These comparisons are easier to see with the difference display* provided in Figure 5.12, which shows the basis of the 7-point overall difference between the 200 mg dose and the placebo. Subtracting the total length of the red bars from the total length of the green bars gives a total difference of 6.7 (rounded up to 7). The 200 mg dose shows five advantages over the placebo. The main advantage for placebo is malignancies.

Difference displays for the 400 mg dose compared to placebo show the same five advantages as the 200 mg dose, but the main comparative advantages of placebo are fewer deaths and fewer cases of tuberculosis. These two placebo advantages are shown in a third difference display to be the advantages of the lower dose over the higher dose, while the higher dose's advantage is a lower proportion of malignancies. These are the only three ways in which the differences between the doses are clinically significant.

5.6.7 Sensitivity Analyses

Participants found assessing the weights expressing the crucial trade-off between the ranges on the ACR70 and deaths scales to be the most difficult task. Would different weights substantially change the overall result? Sensitivity analysis provides an answer: How much can the relative weights of FEs and UFEs differ from the participants' assessments without tipping the balance? Figure 5.13 shows a fairly wide region, defined by the lighter

---

* Obtained with the SORT function of Hiview3.

| Compare | Drug × 200 mg | ▾ | minus | Placebo | | ▾ |

	Model Order	Cum Wt	Diff	Wtd Diff	Sum
FEs	mTSS	18.5	24	4.4	4.4
AEs	SAEs	5.5	72	4.0	8.4
ACR	ACR 50	12.9	29	3.8	12.2
ACR	ACR 20	7.4	47	3.4	15.6
ACR	ACR 70	18.5	16	3.0	18.6
UFEs	Tuberculosis	5.5	−17	−0.9	17.7
UFEs	Deaths	18.5	−9	−1.7	16.0
AEs	Infections	3.9	−78	−3.0	13.0
UFEs	Malignancies	9.2	−69	−6.4	6.7
		100.0		6.7	

FIGURE 5.12 (SEE COLOR INSERT)
Difference display showing the weighted difference in preference scores between the 200 mg dose of drug X compared to placebo, with the effects ordered by the magnitudes of those differences. Right extending (green) bars are for effects that favor the 200 mg option, while left extending bars favor placebo. The "Cum Wt" column gives the cumulative weights (in a different order than shown in the "Cumulative Weight" column shown in Figure 5.11), the "Diff" column shows the difference between the preference values of the options, and the "Wtd Diff" column is the product of the previous two columns. The "Sum" column shows the cumulative sum of the previous column, with the total (6.7) the difference between the total weighted difference scores.

background color, extending from about 17 to 57, within which the UFE node weight can change without affecting the overall result. If the cumulative weight for the UFE node were more than 57 (and therefore the weight on the FE node would be less than 43), neither drug would be preferred to the placebo.

This suggested that the B-R balance might be particularly sensitive to the weight on malignancies. Figure 5.14 confirmed that suspicion: increasing the cumulative weight on malignancies from 9.2 to 12 tipped the B-R balance in favor of the 400 mg dose, though an additional increase beyond about 24 brought the placebo into the most preferred position.

That insight led the group to question sensitivities on the other effects. A quick overall display is shown in Figure 5.15. The weights on three effects are particularly crucial: 400 mg drug X could become the most preferred option with a slight increase in the cumulative weight on either infections or malignancies, or a decrease in the weight on tuberculosis.

The group also developed a convex value function for malignancies, which dropped off sharply from a preference value of 100 for zero malignancies. Using this value function resulted in lower total preference values on all three alternatives and increased the difference between the 200 mg alternative and

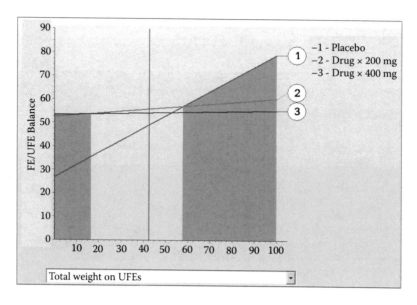

FIGURE 5.13 (SEE COLOR INSERT)
Sensitivity analysis on UFEs. The cumulative weight on UFEs is varied from 0 to 100, while all other weights remain in their original ratios. The vertical red line shows the current value, 42.7, as shown in Figure 5.10. Its intersection with the left–right green, blue, and red lines is at the scores for the three options, 49, 54, and 56. Crossover points are highlighted by the change in the background color.

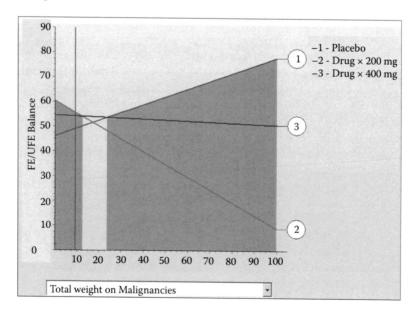

FIGURE 5.14 (SEE COLOR INSERT)
Sensitivity analysis on malignancies, for which the current cumulative weight is 9.2.

the other two. The group used additional individual sensitivity analyses to explore changes on one variable at a time. With the help of software, it was possible to change two weights simultaneously, making it possible to identify the most preferred option for any combination of the two weights.

5.6.8 Scenario Analysis

As participants gained a deeper understanding of the main influences of individual effects on the overall B-R balance, they began to explore combinations of effects that might develop in the future. For example, returning to the red effects shown in Figure 5.15, the group doubled the weights on infections and on malignancies to simulate post-approval concerns about these side effects. Participants also questioned the weight on tuberculosis because the one study showing a high rate of tuberculosis was conducted in a country where that disease was more prevalent. To reflect a more appropriate prevalence, they halved the weight on tuberculosis. The results are shown in Figure 5.16. Under this new set of weights, the 400 mg dose is most preferred, and the difference between the 200 mg dose and placebo is greatly reduced.

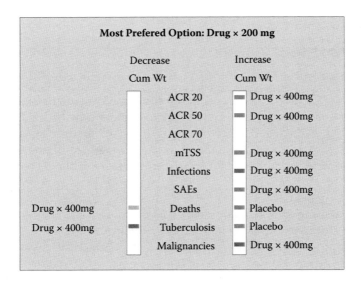

FIGURE 5.15 (SEE COLOR INSERT)
Sensitivity analyses showing a change in the most preferred option, drug X 200 mg, as a result of decreasing (left column) or increasing (right column) the weight on the corresponding effect. A red bar indicates effects for which a change in the weight for that effect alone of less than 5 points is sufficient to shift the most preferred option. A yellow bar signals a preferred option shift if the cumulative weight changes between 5 and 15 points, while a green bar identifies effects requiring more than a 15-point increase. No colored bars indicate that weight changes of any amount do not shift the most preferred option.

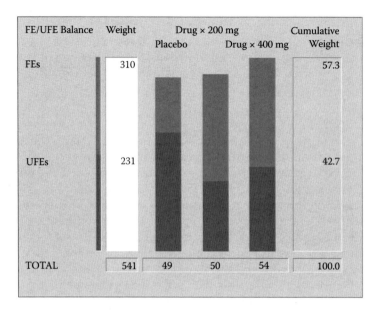

FE/UFE Balance	Weight	Placebo	Drug × 200 mg Drug × 400 mg	Cumulative Weight
FEs	310			57.3
UFEs	231			42.7
TOTAL	541	49 50	54	100.0

FIGURE 5.16 (SEE COLOR INSERT)
Overall result after doubling the weights on infections and malignancies, and halving the weight on tuberculosis. Recall from Figure 5.10 that more green indicates more benefit, and more red indicates greater safety.

5.7 Discussion

After a full exploration, the group found it could recommend that both doses of the drug were approvable. Having met to explore how MCDA could be applied to assess the B-R balance of a drug, the participants concluded that it was helpful to focus on the smaller, more manageable questions of the complex problem and let the MCDA program reassemble the pieces into the larger picture. Participants reported that the group-oriented approach was useful, as it enabled everyone to participate, which surfaced important differences in assumptions and judgments. These differences could be explored through sensitivity analyses, which helped to identify key effects influencing the B-R balance and enabled them to construct recommendations supported by the whole group. However, some of the participants felt that further experience and study would be required before they felt comfortable with the quantitative modeling.

These views are consistent with the findings from the 17 cases of MCDA modeling of the B-R of drugs reported in the public domain: comparison of treatments for children with idiopathic short stature,[36] B-R for rizatriptan,[5] the 4 drugs modeled by master's degree projects of students at the London School of Economics and Political Science,[37–40] 5 new drugs under active consideration by the European Medicines Agency,[20,41] and 6 existing drugs modeled by participants in the IMI's PROTECT project (in press). All but one of these 17 employed MCDA at some point. With proof of principle established, MCDA is emerging as the gold standard for the technical modeling of the B-R balance of drugs.

Establishing the B-R balance of a drug, however, is not just a matter of applying an appropriate technical framework; it also requires a social process to ensure that all relevant perspectives and experiences are brought to bear on the construction of a technical model. Many of the drugs modeled to date employed some sort of facilitated workshop. Drug X employed a decision conference, with a facilitator to guide the group in constructing the model, but not contributing to the content of their discussions. By creating the model on the spot, the decision conferencing process provided structure to the participants' thinking about the issues, encouraged exploration of the effects of the participants' differences of opinion, and enabled the participants to understand the effects of imprecision and uncertainty in the input data (or possible data not yet observed). In helping the participants to construct their preferences,[42] the model acted more like an architect's prototype of a to-be-built structure than as a scientific construct to establish a truth.

Insofar as drugs act differently in different people, and their benefits and risks are consequently perceived differently, there is no right answer for the balance of effects. Instead, there are many answers, all of which can be explored with a requisite model that is fit for purpose, and good enough to make a decision with confidence.[43] A goal of a decision conference is to develop such a model and no more: simple, but not simplistic.

5.7.1 The Double-Counting Bias and Its Correction

The problem of double counting—seen in the three levels of ACR scores for the drug X case and also encountered by the EMA when modeling psoriasis and lupus drugs that use similar nested scores—hints at a larger problem that can induce substantial bias in the intuitive judgment of benefits as well as in an MCDA model. The bias becomes evident in a model once the double counting is removed. Table 5.6 shows the relevant data from Table 5.5 in the left block, and as a frequency distribution in the right block. The three ranges for the ACR scores and the corresponding proportions of patients falling into those mutually exclusive categories ensure that each patient falls into one and only one category.

Entering the revised categories and data into the model gives a rather different result from the totals in Figure 5.10, as can be seen in Table 5.7. The new results show substantially less difference among the options, with only a marginal improvement for either dose of drug X over placebo. Why was this not seen in the sensitivity analysis on the ACR node, one of the many sensitivity analyses performed? Sensitivity analysis showed that the current cumulative weight at the ACR node, the sum of the weights for the three ACR scales, was nearly 40, and would have to be halved before placebo would

TABLE 5.6

ACR Data from Table 5.5 (left) and Those Data as a Frequency Distribution (Right)

	Placebo	200 mg	400 mg		Placebo	200 mg	400 mg
ACR20	11.7	58.2	59.6	ACR20–49	5.9	23.4	23.0
ACR50	5.8	34.8	36.6	ACR50–69	3.4	16.0	20.5
ACR70	2.4	18.8	16.1	ACR70+	2.4	18.8	16.1
				Total	11.7	58.2	59.6

TABLE 5.7

Overall Weighted Preference Values for the Original
Drug X Model, and the New Values When the Double
Counting Is Removed

	Placebo	Drug X 200 mg	Drug X 400 mg
Original	49	56	54
No double counting	48	51	49

become most preferred. That seemed unreasonable to the group, so they felt that the effect of double counting was not serious. It was only later, when the correct structuring of the ACR effects was created, that the bias became clear.

This finding raises a question about the effect of double counting when only intuitive judgment is the basis for determining the B-R balance. Recall the objection that a model wasn't needed because the result was obvious from an effects table, a comment made by an attendee at a workshop where drug X was first presented. Would that comment have been made if the data had been presented as a frequency distribution with mutually exclusive ranges? If not, then a testable working hypothesis is that any display showing a cumulative distribution of frequencies or percentages will bias the perceived benefit effect to appear larger than it actually is—because the same patients appear at different levels of a single effect. That is also true in MCDA modeling, as demonstrated here. In general, the more levels, the greater the bias.

Another bias can be introduced into MCDA models by violating the non-redundancy requirement. Particularly for drugs that treat potentially fatal diseases, mean survival in months is often reported along with mean progression-free survival. Are these really two separate effects? Would one be considered by a patient to be the more fundamental objective,[44] or are they truly two different aspects of survival? They are so closely related that incorporating both in an MCDA model could be considered an inflation of benefits. When both are included in a model, along with double counting of patients in a single effect that can also inflate benefits, a drug with a positive B-R balance, whether judged intuitively or subject to MCDA modeling, can actually tip in the opposite direction when the biases are corrected.*

---

* Personal communication for a drug under development.

The lesson is clear: cumulative data for a given effect must be reframed as a proper frequency (or percentage) distribution, and effect redundancy should be resolved before the related effects are included in an MCDA model. Failing to recognize and correct these biases leaves a question about their effect on decision makers: Does it lead to the continuing development and approval of relatively ineffective drugs?

5.7.2 Why the Resistance?

In many quarters of the pharmaceutical industry and regulatory authorities, substantial resistance is encountered to the prospect of quantitative, explicit modeling of B-R. This resistance needs to be understood, for it is not necessarily irrational. Here are five possible reasons for resistance:

Words are the bricks and mortar of medicine. In the novel by Abraham Verghese,[45] a professor of medicine at Stanford University, the novel's narrator, Marion Stone, is learning medicine by reading medical textbooks (Verghese, p. 268):

> I found that the bricks and mortar of medicine (unlike, say, engineering) were words. You needed only words strung together to describe a structure, to explain how it worked, and to explain what went wrong. The words were unfamiliar, but I could look them up in the medical dictionary, write them down for future use.

> Perhaps this explains why assessment reports are mainly words with some tables of data and perhaps a survival graph or forest plot. There is no need for a quantitative model; assessment can be accomplished with words.

There is no universal scale for balancing benefits and risks; the scale is contingent on culture. The following is a summary in the concluding chapter of Lyn Payer's study of medical practice in France, Germany, the United Kingdom, and the United States.[46] She observes:

> While scientifically conducted studies can show us that a certain course of action or treatment can result in certain benefits and risks, the weighing of those benefits and risks will always be made on a cultural scale.

> How does one decide that, for a given patient, a little exposure to germs is preferable to a sanitized life-style? There is no easy answer, no mathematical formula that can factor risks and benefits and come up with a pat answer as to which treatment should be given and whether treatment should be given at all.

> Neither of those quotes establishes the impossibility of quantitative modeling. Instead, they point out that the B-R balance depends

on value judgments that can differ from one culture to the next. Thus, a drug might be approved in one country but not another, and this can reflect cultural differences in the practice of medicine in those two countries.

Evidence-based decision making in medicine shields the important role of value judgments. Again, the voice of Lyn Payer:

> While medical ethicists and some enlightened doctors are beginning to see the large role value judgments play in medicine and realize that this implies a larger role for the patient in the making of medical decisions, most doctors, of course, continue to hide behind the screen of "scientific" medicine that somehow takes precedence over "unscientific" patient desires.

In one sense, this is the other side of the "It's too subjective" coin. But it is necessary to repeat the distinction between facts and value judgments. Certainly we all wish our doctors to know and be influenced by the evidence that could be brought to bear on a diagnosis and treatment, but we also expect that judgment must also be brought to bear. And when our own judgment is lacking, we may well ask the doctor, "What would you do in this situation?" Judgments will continue to play a significant role in assessing B-R in drug discovery and development, in regulatory practice, and in clinical practice. Quantitative modeling can reveal what those judgments are, and identify which ones are crucial to decision making.

Models are perceived as threatening to the authority of individual doctors. Physicians gain authority by holding the information relevant to a decision. However, decisions are based on preferences, which are formed by information and the decision maker's values. Complete models of B-R make explicit the difference between data and values. Constructing such a model requires making the decision maker's values explicit, which can be seen as diminishing his or her expertise and losing control of the B-R balance. And admitting of the distinction between data and values raises the important question, "Who has the authority to impose values on any medical decision?"

Finding an appropriate role for human judgment. The human brain can keep in mind at one time about seven, plus or minus two, pieces of information, as the psychologist George Miller pointed out over a half-century ago.[47] For most drugs, the B-R balance is characterized by many more than five to nine pieces of information, more than a single assessor can keep in mind when attempting to balance benefits against risks. One European regulator explained* how he deals

---

* Personal communication.

with this complexity: he starts with the benefits, comparing the performance of the drug with the comparator one benefit criterion at a time, building a favorable picture of the drug. He then turns to the side effects, and mentally subtracts these unfavorable effects from the benefits, which leaves him with an overall impression that the B-R balance is either favorable or unfavorable. Contrast this intuitive integration of different sources of data (or any other mental modeling method) with research over the past 60 years comparing unaided human judgment with simple linear models. Meehl[12] has been consistent in identifying the problem as the integration of multiple pieces of data, not the judgments about the pieces, a finding confirmed in nonclinical settings by the extensive research showing that people are slow to update their uncertainty as they receive more and more information.[48–50] In those studies, a simple linear model for combining the data, based on Bayes' theorem, outperformed the assessors in combining their own judgments about individual pieces of data.

The implications of these findings for B-R assessment is that an assessor may focus on only a few critical effects, and will require more information and take longer to come to a final judgment than is necessary. However, judgment is still needed at every one of the eight steps shown in Table 5.4 for constructing an MCDA model (except for step 6, when the task of putting data and judgments together so the resulting B-R balance can be explored under different assumptions is delegated to a computer). In a sense, the model hands back that information in changed form, which can provide new perspectives on a problem and trigger insights that were not obvious from inspection of the input information. The properties of water are not at all obvious from study of the separate properties of hydrogen and oxygen.

Appropriate application of decision-analytic models, with help from some of the other models under the middle-column approaches of Table 5.2, could pave a path forward for moving decision making about the B-R balance of medicinal products from implicit to explicit and from qualitative to quantitative.

5.8 Conclusions

Engaging all key players in constructing, face-to-face, a quantitative model of the B-R balance of drugs is now beyond the proof-of-concept stage. Whether or not it is desirable to move in this direction has yet to be demonstrated. Fortunately, applying decision-theoretic approaches, based on the simple mathematical models of expected utility and weighted utility, is not particularly difficult or time-consuming. All five models created for new, as-yet-unapproved drugs under active consideration by the Committee for Medicinal Products for Human Use (CHMP) were created within 6 hours in a facilitated workshop attended by knowledgeable experts and regulators, who shared their knowledge and experience. The process helped them to construct their preferences and, with iteration through the MCDA steps of Table 5.4, develop a shared understanding of the B-R balance that was sufficient to formulate a recommendation. This is how quantitative modeling can enhance, but by no means replace, the capability of decision makers to judge whether the B-R balance of a medicinal product is fit for purpose.

References

1. Eichler, H.-G., Abadie, E., Raine, J.M., and Salmonson, T. 2009. Safe drugs and the cost of good intentions. *New England Journal of Medicine* 360: 1378–1380.
2. Phillips, L.D. 1993. Decision theory and its relevance to pharmaceutical medicine. In *Textbook of pharmaceutical medicine*, ed. R.D. Mann, M.D. Rawlins, and R.M. Auty, 247–255. Carnforth, Lancashire: Parthenon Publishing Group.
3. CIOMS IV. 1999. *Benefit-risk balance for marketed drugs. Evaluating safety signals. Geneva*: Council for International Organizations of Medical Sciences.
4. Mussen, F., Salek, S., and Walker, S. 2007. A quantitative approach to benefit-risk assessment of medicines. Part 1. The development of a new model using multi-criteria decision analysis. *Pharmacoepidemiology and Drug Safety* 16: S2–S15.
5. Mussen, F., Salek, S., and Walker, S. 2009. *Benefit-risk appraisal of medicines: A systematic approach to decision-making*. Chichester: John Wiley & Sons.
6. European Medicines Agency. 2012. *Benefit-risk methodology project: Report on risk perception study module*, 1–68. London. www.ema.europa.eu, Special topics, Benefit-risk methodology.
7. Wallsten, T., Budescu, D., and Zwick, R. 1993. Comparing the calibration and coherence of numerical and verbal probability judgments. *Management Science* 39: 176–190.
8. Kahneman, D. 2011. *Thinking, fast and slow*. London: Allen Lane.
9. Raiffa, H. 1968. *Decision analysis*. Reading, MA: Addison-Wesley.
10. Keeney, R., and Raiffa, H. 1976. *Decisions with multiple objectives: Preferences and value tradeoffs*. New York: John Wiley. (Republished 1993, Cambridge University Press.)
11. Weinstein, M., Torrance, G., and McGuire, A. 2009. QALYs: The basics. *Value in Health* 12: S5–S9.
12. Meehl, P. 1996. *Clinical versus statistical prediction: A theoretical analysis and a review of the evidence*. Northvale, NJ: Jason Aronson. (Original 1954 edition, University of Minnesota Press.)
13. Grove, W., and Meehl, P. 1996. Comparative efficiency of informal (subjective, impressionistic) and formal (mechanical, algorithmic) prediction procedures: The clinical-statistical controversy. *Psychology, Public Policy, and Law* 2: 293–323.

14. European Medicines Agency. 2009. Work package 1 report: *Description of the current practice of benefit-risk assessment for centralised procedure products in the EU regulatory network*, 23. London. www.ema.europa.eu, Special topics, Benefit-risk methodology.

15. European Medicines Agency. 2010. *Guidance document: Day 80 critical assessment report*.

16. Guo, J., Pandey, S., Doyle, J., Bian, B., Lis, Y., and Raisch, D. 2010. A review of quantitative risk-benefit methodologies for assessing drug safety and efficacy: Report of the ISPOR risk-benefit management working group. *Value Health* 13: 657–666.

17. Holden, W. 2003. Benefit-risk analysis: A brief review and proposed quantitative approaches. *Drug Safety* 26: 853–862.

18. European Medicines Agency. 2010. *Work package 2 report: Applicability of current tools and processes for regulatory benefit-risk assessment*, 33. London. www.ema.europa.eu, Special topics, Benefit-risk methodology.

19. Shahrul M., et al. 2012. *Benefit-risk integration and representation*. London: PROTECT Consortium.

20. European Medicines Agency. 2011. *Work package 3 report: Field tests*, 29. London. www.ema.europa.eu, Special topics, Benefit-risk methodology.

21. Belton, V., and Stewart, T. 2002. *Multiple criteria decision analysis: An integrated approach*. Boston: Kluwer Academic Publishers.

22. Dodgson, J., Spackman, M., Pearman, A., and Phillips, L. 2000. *Multi-criteria analysis: A manual*. London: Department of the Environment, Transport and the Regions. (Republished 2009 by the Department for Communities and Local Government.)

23. Ramsey, F. 1926. Truth and probability. In *The foundations of mathematics and other logical essays*, ed. R. Braithwaite, 156–198. London: Kegan, Paul, Trench, Trubner & Co.

24. von Neumann, J., and Morgenstern, O. 1947. *Theory of games and economic behavior*. 2nd ed. Princeton, NJ: Princeton University Press.

25. Savage, L. 1954. *The foundations of statistics*. New York: Wiley. (2nd ed., 1972, New York: Dover Publications.)

26. Raiffa, H., and Schlaifer, R. 1961. *Applied statistical decision theory*. Cambridge, MA: Harvard University Press.

27. Howard, R. 1966. Decision analysis: Applied decision theory. In *Proceedings of the Fourth International Conference on Operational Research*, ed. D. Hertz and J. Melese, 55–71. New York: Wiley-Interscience.

28. Hammond, J., Keeney, R., and Raiffa, H. 1999. *Smart choices: A practical guide to making better decisions*. Boston: Harvard University Press.

29. Phillips, L. 2007. Decision conferencing. In *Advances in decision analysis: From foundations to applications*, ed. W. Edwards, R. Miles, and D. von Winterfeldt, 375–399. Cambridge: Cambridge University Press.

30. Franco, L., and Montibeller, G. 2010. Facilitated modelling in operational research. *European Journal of Operational Research* 205: 489–400.

31. http://en.wikipedia.org/wiki/Hiview3.

32. http://www.visadecisions.com/.

33. http://www.logicaldecisions.com/.

34. http://www.m-macbeth.com/en/m-home.html.

35. Walker, S., Phillips, L., and Cone, M. 2006. *Benefit-risk assessment model for medicines: Developing a structured approach to decision making.* Epsom: Centre for Medicines Research International, Institute for Regulatory Science.
36. Felli, J., Noel, R., and Cavazzoni, P. 2009. A multiattribute model for evaluating the benefit-risk profiles of treatment alternatives. *Medical Decision Making* 29: 104–115.
37. Alexiev, V. 2010. *Modelling the balance of clinical efficacy and safety of drugs in the context of regulatory evaluation for marketing authorization.* London: London School of Economics and Political Science.
38. Ali-Basah, C. 2010. *Modelling the balance of favourable and unfavourable effects of a drug in the presence of uncertainty for a regulatory environment: The case for Cimzia using multi criteria decision analysis (MCDA) and Monte Carlo simulation.* London: London School of Economics and Political Science.
39. Kochanowska, E. 2010. *Modelling the balance of favourable and unfavourable effects of a drug in the presence of uncertainty: The case of Acomplia with the use of multi-criteria decision analysis.* London: London School of Economics and Political Science.
40. Thoi, D. 2010. *Quantitative risk-benefit analysis of Sutent using a Markov model.* London: London School of Economics and Political Science.
41. European Medicines Agency. 2012. *Work package 4 report: Benefit-risk tools and processes,* 20. London. www.ema.europa.eu, Special topics, Benefit-risk methodology.
42. Lichtenstein, S., and Slovic, P., eds. 2006. *The construction of preference.* New York: Cambridge.
43. Phillips, L. 1984. A theory of requisite decision models. *Acta Psychologica* 56: 29–48.
44. Keeney, R. 1992. *Value-focused thinking: A path to creative decisionmaking.* Cambridge, MA: Harvard University Press.
45. Verghese, A. 2009. *Cutting for stone.* New York: Vintage Books.
46. Payer, L. 1996. *Medicine and culture.* New York: Henry Holt and Company.
47. Miller, G. 1956. The magical number seven, plus or minus two: Some limits on our capacity for processing information. *Psychological Review* 63: 81–97.
48. Edwards, W. 1968. Conservatism in human information processing. In *Formal representations of human judgment,* ed. B. Kleinmuntz, 17–52. New York: Wiley.
49. Edwards, W., Phillips, L., Hays, W., and Goodman, B. 1968. Probabilistic information processing systems: Design and evaluation. *IEEE Transactions on Systems Science and Cybernetics* SSR-4; 248–265.
50. Phillips, L., Hays, W., and Edwards, W. 1966. Conservatism in complex probabilistic inference. *IEEE Transactions on Human Factors in Electronics* HFE-7: 7–18.

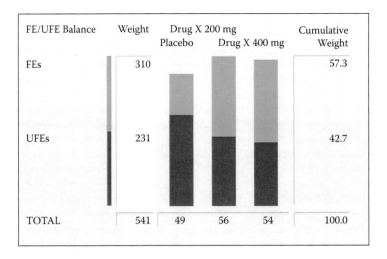

FIGURE 5.10
Overall results are shown in the "Total" row. Longer upper (green) sections mean more benefit, while longer lower (red) sections indicate more safety. The two numbers in the "Weight" column are the sums of the nonnormalized input weights (terminal node products from Figure 5.6) for the FEs and UFEs, while the "Cumulative Weight" column shows the normalized weights.

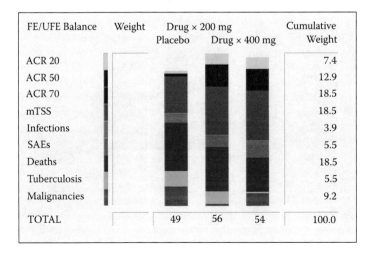

FIGURE 5.11
The contributions of each effect to the overall weighted preference values.

	Model Order	Cum Wt	Diff	Wtd Diff	Sum	
FEs	mTSS	18.5	24	4.4	4.4	
AEs	SAEs	5.5	72	4.0	8.4	
ACR	ACR 50	12.9	29	3.8	12.2	
ACR	ACR 20	7.4	47	3.4	15.6	
ACR	ACR 70	18.5	16	3.0	18.6	
UFEs	Tuberculosis	5.5	−17	−0.9	17.7	
UFEs	Deaths	18.5	−9	−1.7	16.0	
AEs	Infections	3.9	−78	−3.0	13.0	
UFEs	Malignancies	9.2	−69	−6.4	6.7	
		100.0		6.7		

Compare [Drug × 200 mg ▾] minus [Placebo ▾]

FIGURE 5.12

Difference display showing the weighted difference in preference scores between the 200 mg dose of drug X compared to placebo, with the effects ordered by the magnitudes of those differences. Right extending (green) bars are for effects that favor the 200 mg option, while left extending bars favor placebo. The "Cum Wt" column gives the cumulative weights (in a different order than shown in the "Cumulative Weight" column shown in Figure 5.11), the "Diff" column shows the difference between the preference values of the options, and the "Wtd Diff" column is the product of the previous two columns. The "Sum" column shows the cumulative sum of the previous column, with the total (6.7) the difference between the total weighted difference scores.

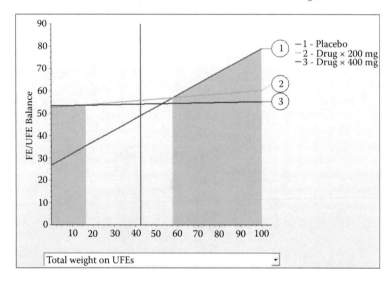

FIGURE 5.13

Sensitivity analysis on UFEs. The cumulative weight on UFEs is varied from 0 to 100, while all other weights remain in their original ratios. The vertical red line shows the current value, 42.7, as shown in Figure 5.10. Its intersection with the left–right green, blue, and red lines is at the scores for the three options, 49, 54, and 56. Crossover points are highlighted by the change in the background color.

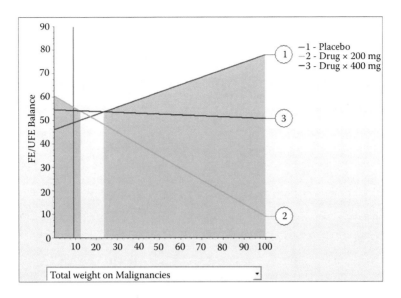

FIGURE 5.14
Sensitivity analysis on malignancies, for which the current cumulative weight is 9.2.

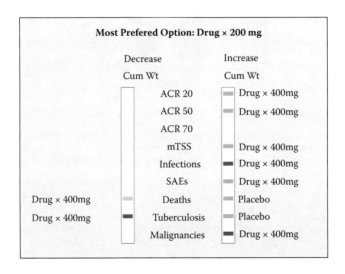

FIGURE 5.15
Sensitivity analyses showing a change in the most preferred option, drug X 200 mg, as a result of decreasing (left column) or increasing (right column) the weight on the corresponding effect. A red bar indicates effects for which a change in the weight for that effect alone of less than 5 points is sufficient to shift the most preferred option. A yellow bar signals a preferred option shift if the cumulative weight changes between 5 and 15 points, while a green bar identifies effects requiring more than a 15-point increase. No colored bars indicate that weight changes of any amount do not shift the most preferred option.

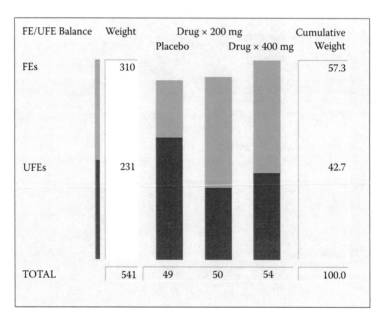

FIGURE 5.16
Overall result after doubling the weights on infections and malignancies, and halving the weight on tuberculosis. Recall from Figure 5.10 that more green indicates more benefit, and more red indicates greater safety.

6

Benefit-Risk Communication: Learning from Our Past and Creating Our Future

Marilyn Metcalf

CONTENTS

6.1 The Importance of Communication

We have seen in previous chapters the central role of communication in benefit-risk (B-R) evaluation. B-R assessments are, at their heart, structured exchanges of information. While there is potential for bias in the ways questions are asked and recommendations are framed, the methods described in this book demonstrate techniques for addressing bias and conducting B-R assessments more objectively. In addition to this analytical component, part of the function of a B-R assessment is to draw upon and incorporate human judgment. This evaluative aspect makes appropriate communication around the evidence for benefit and risk all the more important. Utilized appropriately, B-R frameworks can provide coherent ways to contemplate different goals, values, and perspectives in healthcare decisions. The following examples of medical issues seen through a B-R lens describe experiences that can be instructive in considering directions for the future development of B-R communications.

6.2 Example 1: Seeking Commonality

In the early days of the spread of the human immunodeficiency virus (HIV) in the West, an activist group of HIV patients increasingly voiced their need for treatments. The B-R question revolved around the level of evidence needed to approve a drug for treatment of this deadly, infectious disease. The benefit of effective treatment was prolonged life (by a matter of months in the beginning). The risks of treatment were the possibility that inadequately tested drugs might carry side effects that would make patients even sicker, and the chance that suboptimal treatment might allow the virus to mutate faster and become more difficult to treat. A documentary filmmaker described the initial efforts of ACT UP (AIDS Coalition to Unleash Power) in 1987 as emotional appeals from a mostly gay, mostly male group of patients who were living with what came to be called a plague. When drugs were not available in the United States, the group began importing drugs that were approved in other countries. Inexplicably, a white, middle-aged, straight woman started to attend meetings and attempt to speak to the group. She seemed very out of place and had difficulty being heard until a member of ACT UP encouraged the group to listen. The woman was a retired pharmaceutical chemist who wanted to help the group find treatments, so she gave them insights into the workings of drug development and approval. HIV patients began to study the science of HIV infection, to attend medical conferences not just as protesters but as participants, and challenged the healthcare field. This community of patients and caregivers recognized the need to learn practices and language that were foreign to them so that they could work effectively with the established systems that develop, review, and approve medicines.[1,2]

For their part, healthcare providers, pharmaceutical companies, and regulators had to open the doors of their medical conferences and agencies to laypeople. They began to work with vocal advocate organizations and learn about communities of patients who were actively discussing clinical trials and alternative treatments among themselves. Time allotted for patients to share personal stories at AIDS conferences increasingly included discussions about therapies, clinical trials, alternative medicines, drug regimens and drug holidays, quality of life, surrogate measures of efficacy, and survival. Regulatory processes could be accelerated once there was sufficient understanding of the disease to agree upon useful and acceptable surrogate markers. All of these topics were of interest to all engaged parties, but common language and common understanding enabled the clarity and focus that made the communications and subsequent efforts productive.

As the filmmaker described it, "When they finally all got together, and we see that in the course of this film, what they discovered was, you know, real human beings trying as hard as they could on all sides to do the right thing, and that in fact the scientists were trying to be heroic, and the AIDS activists were trying to be helpful. And once they made that bond ... the beginning of that dialogue and respect and trust, then they really started to pull together in a single unit to try to find their way through this morass of this new virus, and to try to find a way to bring it to its knees."[2]

The AIDS crisis is by no means over, and many challenges and questions around B-R, treatment, prevention, and a cure remain. But this small anecdote, one in a vast number of global efforts, teaches important lessons about having common goals and a common language, and the value to be gained by sharing spaces that were once the exclusive domain of a single group. Useful B-R frameworks include not only structure for outputs such as recommendations, but also processes that gather information dynamically from a variety of perspectives. There is value in communication among people who may not be accustomed to working together, as they recognize their mutual interests and converge upon common goals. The act of framing and creating a B-R evaluation is the first area of B-R communication, and is vital in setting the stage for all other communications about benefit and risk.

6.3 Example 2: Finding Clarity

The second example involves a much broader and more diffuse population of stakeholders. We will focus on the communications in two countries in which the popular press became involved in publicizing the B-R questions. This example begins in the UK and widens to include the United States. The benefit-risk question was over the value of mammography for lowering the mortality rate from breast cancer in women.[*] Does mammography offer more benefit by detecting cancers that would not otherwise have been detected and thereby lower mortality? Or does mammography cause more harm by detecting benign conditions that are then treated (overtreated), potentially causing morbidity to women who otherwise would not have been deemed to need treatment because they would not have developed breast cancer? Worse still, does mammography cause this morbidity and fail to lower mortality?

A public debate started in earnest in 2009 with an open letter[4] from researchers to the *Times* London newspaper, challenging the United Kingdom's National Health Service (NHS) information pamphlet on breast cancer screening.[5] The *Times* referenced a *BMJ*[†] article that was being published on the same topic,[6] as well as a testimonial from a woman who felt that she had been overtreated and had not been provided sufficient opportunity to participate in her healthcare process.

The NHS pamphlet of concern explained what mammography was, how and when it was done, and who was eligible. "To help you decide whether or not to come for breast screening," it characterized the "main benefits and difficulties" as follows:

---

[*] Note: Men can develop breast cancer, but it is so rare that mammography among the general population of men is not recommended as a screening approach, so this example specifies women as the patient population.[3]

[†] The *British Medical Journal* officially changed its name to *BMJ* in 1988.

- Most breast cancers are found at an early stage when there is a good chance of a successful recovery.
- Around half the cancers that are found at screening are still small enough to be removed from the breast. This means that the whole breast does not have to be removed.
- Breast screening saves an estimated 1,400 lives each year in this country.
- Breast screening reduces the risk of the women who attend dying from breast cancer.
- We will call back some women for more investigations if we are not sure about their mammogram. After more tests, we will find that many of these women will not have cancer. If you are called back it can cause worry.
- Screening may miss some breast cancers.
- Not all breast cancers that are found at screening can be cured.
- Many women find mammography uncomfortable or painful, but normally just for a brief period of time.

A proposed alternative pamphlet summarized the benefits and risks in this way. "If 2000 women are screened regularly for 10 years, one will benefit from the screening, as she will avoid dying from breast cancer. At the same time, 10 healthy women will, as a consequence, become cancer patients and will be treated unnecessarily. These women will have either a part of their breast or the whole breast removed, and they will often receive radiotherapy, and sometimes chemotherapy. Furthermore, about 200 healthy women will experience a false alarm. The psychological strain until one knows whether or not it was cancer, and even afterwards, can be severe."[7]

Farther down in the alternative pamphlet was an elaboration on the effects of overdiagnosis and overtreatment. For "false alarm," it explained, "if 2000 women are screened regularly for 10 years, about 200 healthy women will experience a false alarm. The psychological strain until it is known whether or not there is a cancer can be severe. Many women experience anxiety, worry, despondency, sleeping problems, changes in the relationships with family, friends and acquaintances, and a change in sex drive. This can go on for months, and in the long term some women will feel more vulnerable about disease and will see a doctor more often." Included were summaries of clinical trial results to document the effect on mortality, the numbers of women who experienced false alarms, and the number of women who reported pain associated with the mammogram. No further information was provided on anxiety, worry, despondency, sleep problems, changes in relationships, or changes in sex drive.

The *BMJ* article referenced in the letter to the newspaper included a citation that appeared to be the source material for the evidence of psychological strain. When traced, it was one author's PhD thesis[8] plus a related article[9] reviewing 23 studies and validating a new research instrument on psychosocial consequences of false positives. Going back to that source, "Given the inadequacy of the measurement instruments used in previous quantitative studies on psychosocial aspects of false positives, any current conclusions about the psychosocial long-term consequences of false positives must remain tentative."[8] Two weeks after an abnormal screening mammography the negative consequences were said to decrease, but remain of measurable significance. One to six months after a false positive screening, women showed changed attitudes toward what were termed "existential values." These included attitudes toward valuing life, enjoying life, feeling more optimistic or pessimistic about the future, as well as changes in relationships within their social network, feeling more or less relaxed and calm, and either more or less confident or anxious about having breast cancer. All of these changes were regarded as harms. "The changes after the false-positive result can only be regarded as effects of a fear induced by the abnormal screening mammography and later the relief induced by the final false-positive result. It is difficult to regard these changes as anything but harm caused by the screening programme."[8]

A related article in the *New York Times* included quotes from a number of experts on each side of the argument, who either favored current breast cancer screening practices or favored changing the wording of the recommendations. When questioned about the statistical arguments against mammography, the director of cancer screening programs for the British National Health Service at the time said the patient handout was being revised and noted that information about overdiagnosis might be added, but dismissed the figures cited in *BMJ* as inaccurate. British studies, she said, showed that the ratio of lives saved to lives unnecessarily disrupted was more like one to one. "We know, from statistics, that there are cancers diagnosed through screening that wouldn't otherwise have been diagnosed—because the woman dies of something else first … or she might live to 90 and it would just sit there, and she wouldn't have died of breast cancer," but "you don't know who that woman is."[10]

In going back to the source articles for the statistics around mammography effectiveness, the reason for disagreement seemed to stem from shortcomings in the available data that made it very difficult to answer the question definitively. Part of the difficulty in determining benefit was finding consistent definitions of precancerous and possibly precancerous conditions, as well as understanding abnormal findings on mammography and their positive predictive value for malignancy. There were also difficulties comparing findings across studies, in part due to differences in classification and inconsistent nomenclature. These findings helped to elucidate the reasons for unresolved debate, but unfortunately did not solve the immediate problem of helping to decide whether benefit outweighed risk. There remained research to be done to answer fundamentally important questions about the natural history of breast cancer. All parties involved seemed to agree that mammograms detected more conditions, including benign abnormalities and potentially precancerous findings, than would have been detected without mammography. Medical sources also suggested that the characterization of mammography results required skill, and reviews of study findings would be better served if characterization could be done in more consistent ways.[11]

The research at that time concluded that the currently available evidence supported the efficacy of screening 50- to 69-year-old women by mammography for reducing mortality from breast cancer. It also confirmed that there was limited evidence for the efficacy of screening 40- to 49-year-old women by mammography. The research reviewed in the publications concluded that there was some benefit to screening, but left open the question of how to characterize the risk and how to weigh the risk against the benefit of screening. When the Preventive Services Task Force of the Department of Health and Human Services in the United States issued new screening guidelines in line with the research, however, there was immediate concern. The guidelines recommended against routine screening mammography for women aged 40 to 49, and scaling back screening for women aged 50 to 74 from annually to every other year. They noted the unclear benefit of mammography for women 75 and older, and suggested not teaching women to perform breast self-exams—nor could they support the usefulness of clinical breast exams.[12]

Reactions to the guidelines were consistent with what behavioral economists and psychologists predicted.[13] Practitioners and patients based their responses on their experiences. A hospital director said she would not change her recommendations about early screening. "I can't tell you how many times we pick up a cancer in a young woman who comes in for a baseline mammogram before age 40.... We are here in the trenches, and we see so many patients dealing with this diagnosis, and we're always happy for the ones who do come in early.... I feel like their life is changed in a better way." Another doctor saw the difficulties of women who frequently had false positives and hearkened back to the psychological stress described above. "Some women are constantly being called back.... It's like a revolving door, in and out, every three to six months, year in and year out. That is insane, and it turns out to be nothing." Likewise, patients who had been treated for breast cancer or whose friends had been treated when they were in their forties were concerned about the dangers of not screening early enough. Women who did not refer to these experiences and who found mammograms uncomfortable described their relief at being asked to have the test later and less frequently.[12]

The debate continued in subsequent years, with very little change in the conclusions. A more recent U.S. study of the impact of 30 years (1976 through 2008) of mammography screening on the stage-specific incidence of breast cancer (i.e., looking to see if breast cancer is caught earlier since screening began as opposed to before screening) suggested that despite the increase in number of early-stage breast cancer cases detected, screening mammography only marginally reduced the rate at which women presented with advanced cancer, which the researchers concluded would suggest substantial overdiagnosis. They added:

> And although no one can say with certainty which women have cancers that are over-diagnosed, there is certainty about what happens to them: they undergo surgery, radiation therapy, hormonal therapy for 5 years or more, chemotherapy, or (usually) a combination of these treatments for abnormalities that otherwise would not have caused illness. Proponents of screening should provide women with data from a randomized screening trial that reflects improvements in current therapy and includes strategies to mitigate over-diagnosis in the intervention group. Women should recognize that our study does not answer the question "Should I be screened for breast cancer?" However, they can rest assured that the question has more than one right answer.[14]

The communications around mammography play a different role than the communications described in the first example involving HIV. In retrospect, the first example was similar to construction of a B-R evaluation using a framework—various perspectives were brought together to discuss common goals. This second communication around mammography is more similar to what used to be called simply risk communication, in which the results of a (benefit-) risk analysis are being reported and recommendations are being made, or in this case, challenged. One lesson to be learned from this latter example is the importance of discussing the limitations of data, and appropriately distinguishing information that analyses can and cannot support.

The available data on mammography have been gathered for different purposes, using various methods, from a variety of patients. As is common with secondary analyses, there is some "noise" in the data, which can make it more difficult to interpret. A thorough analysis of information from a variety of sources necessitates research to understand the data, the methods of data collection, selection of subjects, and other characteristics of the studies. Often, a number of criteria are used to distinguish which sources are the most reliable for the purposes at hand. For the *BMJ* article, a systematic review of the sources for the statistical claims around the effectiveness of mammography in lowering mortality was conducted. There was less research cited around the claims for psychological harm.

The Behavioral Risk Factor Surveillance System (BRFSS) was the source of information on patients' frequency of obtaining mammograms in the more recent 30-year overview of mammography effects on stage-specific breast cancer incidence. This is self-reported data from the U.S. Centers for Disease Control and Prevention (CDC). The sample of patients surveyed regarding their mammography screenings was selected to be representative, and responsible research practices were described for collecting the data. The data carried with them a footnote stating that the questions concerning use of mammography differed slightly over the years in question. A study posted on the CDC Web site suggested that wording and other factors may have overestimated mammography use, and more so for black women than for white women.[15] It would be specious to suggest whether or how much this difference in estimated mammography use might change the findings in the recent mammography effectiveness study; however, it should be useful to walk through the logic of how B-R communications might frame the issue to bring in a number of perspectives and begin to examine the data and their implications.

The researchers at CDC looking at the implications of reporting discrepancies in self-reported data were seeking the answer to a long-time conundrum. There is a "breast cancer incidence-mortality paradox" that shows death rates from breast cancer are higher for black women than for white women, even though breast cancer incidence is higher among white women. The researchers stated their belief that mammograms are "a key tool for detecting breast cancer at an early, treatable stage and thereby reducing death rates from the disease." They were looking for evidence of why black women are more likely than white women to have advanced-stage cancer at diagnosis when data indicate that black women receive mammograms as frequently as white women. They found that studies focusing on racial, ethnic, and socioeconomic disparities did not show a disparity in mammography use between black and white women. And yet, women of lower socioeconomic status, immigrant women, black women, and women from other minority racial or ethnic groups show disparities in stage of breast cancer at diagnosis and breast cancer death rates. Why? One explanation, or partial explanation, could be overreporting of mammography use, as suggested by a meta-analysis of women's self-reported mammography use versus a confirmatory review of medical or billing records.[16] Another (partial) explanation may be in the collection of the information. The question respondents normally answered was preceded by: "These next questions are about mammograms, which are X-ray tests of the breast to look for cancer." In 1992, the BRFSS survey used different introductory wording: "I would like to ask you a few questions about a medical exam called a mammogram. A mammogram is an X-ray of the breast and involves pressing the breast between two plastic plates." The reported prevalence of mammogram use was lower when this more explanatory wording was used, and the reduction was greater among black women than among white women.[15]

The researchers looking at racial disparity acknowledged that there may be differences in the biology of the disease between black women and white women, differences in stage at diagnosis that are unassociated with mammography use, and differences in treatment following diagnosis. They also suggested that the reported black-white parity in mammography use should be reexamined. Their findings showed the percentage of women aged 40 years or older in 1995 who reported receiving a mammogram during the previous 2 years, when adjusting for the potential impact of overreporting, was 54% among white women and 41% among black women, compared with 70%

among both white and black women after the adjustment normally made for age only. In 2006, the percentage receiving mammograms, after adjustment for potential overreporting, was 65% among white women and 59% among black women, compared with 77% among white women and 78% among black women after adjustment for age only.

While there is a clear link between communication concerning the validation of data collection and respondent comprehension of questions, it is not a specific B-R issue. The B-R communication piece comes in the broader presentation of the varieties of data and explanations of the potential pitfalls in using multiple data sources. As camps formed over the question of the value that mammography may or may not add versus the harm that may or may not result, and for whom, there were holes in the information that needed to be noted and addressed as part of B-R assessment and communication. Determining an appropriate venue and level of detail for wide discussion would help to facilitate more informative communications. The public discussion of this topic in the popular press allowed for communication with a wide audience, likely to be far more inclusive than keeping the discussion within scientific journals, but it was not conducive to in-depth discussions of the data, data sources, and data collection methodologies. In addition, as this example illustrates, potentially relevant research might not be seen or discussed unless researchers focused on different but related research questions can be brought together more intentionally (e.g., to explore whether the impact of race or class has any bearing on the impact of screening on stage of diagnosis). Broader inclusion of multiple researchers, healthcare providers, and patients in a larger B-R communication effort, geared toward a more dynamic communication process, might encourage a more mutual goal regarding informing the public of evidence and uncertainty around important public health questions. Ideally, there might be an opportunity for interested parties to contribute to a more communal understanding of conclusions about which there is more confidence as well as areas needing further research. Potential venues for these discussions may include patient-centered efforts or patient-inclusive organizations and conferences that have begun to address B-R questions for disease areas or issues of efficacy, safety, and effectiveness of medicines.[17–22]

6.4 Informing Decisions

Day-to-day healthcare decisions still must be made in the context of open questions around study data and disease progression. Arguably, better choices could be made with more direct information. It may be edifying to consider the perspective shared by the patient who was quoted in the *Times* article about mammograms. She expressed her frustration that at the time of her treatment, she felt that she had little say in her care as she was moved through the healthcare system. We have already noted that in recent years patients have gained an increasing voice in both their own healthcare and study- and population-level discussions regarding medicine. As proposed in earlier chapters, it is possible that providing additional information about treatment alternatives and testing procedures for patients to make more informed choices will address some of the remaining concerns. But uncertainty remains regarding the health impact for specific individuals. We do not have the answers at this point concerning which women have cancer and which are overdiagnosed (Example 2), or how to balance the concerns and risk tolerance of individual patients with the public health concerns and risk tolerance of healthcare professionals (Example 1). These are only a few of the myriad issues that linger in a complex healthcare context, and they illustrate the overarching question for patients and their healthcare providers: How does all of this information help the individual?

In Example 1, the experience with HIV patients demonstrated the value of broad-based communication and inclusion of patient communities in providing information for both patients and the healthcare industry. Articulation of common goals and a willingness to empathize with the perspectives of others enabled more directed partnerships in considering drug regimens, clinical trials, patterns of transmission, and approaches to decision making regarding treatment. It must be acknowledged that those interactions occurred with a small subset of the patient population, and some of the intense focus among the interested parties grew from a crisis. Nonetheless, the experiences suggest that a structured approach characterized by B-R goal identification and analytical thinking could enable productive communication around other health issues. Working with other stakeholders toward mutual benefits seems at face value to provide an opportunity to make more informed decisions and for patients to work more closely with their healthcare providers in their individual care.

What about the more diverse circumstances of women in Example 2? The broad nature of the problem and the diversity of data suggest that communication could be appropriately focused using a B-R framing approach. For example, most of the literature cited focused on one potential benefit of mammography: a decrease in overall mortality. Are there also benefits around morbidity or other treatment effects? Furthermore, clarity is needed regarding the goal of broad, essentially one-way B-R communications, such as publication of research results. Is the goal of communication to raise general awareness? To make recommendations for public health? To seek response from interested parties? Or something else? Is there an embedded or assumed B-R goal of informing individual healthcare decisions, and if so, how is that to be accomplished?

In the mammography example, does any of the information provided in the newspaper article or the pamphlets answer the questions of an individual woman trying to decide whether or not to attend her mammogram appointment? If current research is not designed to inform an individual patient's choices, consideration needs to be given to what data and what findings will inform patients and their healthcare providers. While every patient may differ in her priorities and the combinations of risk factors she must consider, it seems reasonable to try to outline ways in which research can address areas of immediate concern, with acknowledgment that more work is needed to gather additional information and define the most appropriate ways of incorporating multiple viewpoints in decision making. For example, we should not lose sight of the message that research continues to confirm the benefits of mammography for women aged 50–74 years old, with some benefits for 40–49 year olds as well. To examine the risks, healthcare providers and patients need sufficient information to compare the probability of unnecessary surgery against the possibility of malignancy in the future. They also require information regarding what the surgery entails. Do most women have lumpectomies or mastectomies? Under what circumstances? What happens if chemotherapy or radiation is recommended? How long do side effects generally last? What is the risk of cardiovascular harm from radiation? There are many questions that must be addressed to inform what on the surface appears to be a single, relatively simple question.

In addition, patient experiences have changed over the years since mammography's introduction and subsequent widespread use. It may therefore be necessary to reexamine its associated risks. The literature acknowledged improvements in treatment for (suspected) breast cancer; there have also been advances in breast cancer biopsies. In some cases, partial removal of the breast has been reduced to minimally invasive guided needle biopsy with a typical recovery time of 24 hours. Four-fifths of these biopsies come back negative for cancer. It may be helpful for interested parties to have the opportunity to discuss differing perspectives around screening and treatment alternatives. Is it unnecessarily worrisome to use needle biopsies in the way they are used now if 80% of women who receive them need no further treatment, or is it efficient to have a minimally invasive procedure that allows focus on the 20% of those women whose biopsy results need further evaluation with more invasive procedures? These are the kinds of questions that need to be added for a full discussion of the B-R balance of screening procedures and further treatment.[23]

6.5 Assessing Benefit-Risk over the Treatment Life Cycle

As noted in the HIV example, there are public health concerns around B-R that should be considered in communications. At the population level, the B-R of treatments is increasingly being followed over time to ensure that the B-R balance remains positive. After medicines and other therapies are approved and become available to wider patient populations, large studies are sometimes undertaken to answer specific questions, but most B-R information is collected from retrospective looks at spontaneous reports of patient experiences and from monitoring literature or observational data. These sources add incrementally to the information already known. Updates of information on products are reported to regulators, and issues of public concern are more widely reported. Increasingly, regulators are requiring that the B-R balance be explicitly addressed in these updates.

As risk communication has become an articulation of benefit and risk in context of one another, there is growing acknowledgment that there is not just one objective truth to be found in B-R. There is evidence to be gathered and weighed, and judgments to be acknowledged and discussed.[1,22] There is also expertise and experience to consider: the availability of specialist and technical knowledge to a broad public, interspersed with less than accurate information to be sorted through, has widened considerably the conversation around healthcare. Communication about medicines evolves with these sources of information, and the conversation is beginning to include many more voices. What does that broader communication look like, and what does it mean for the future?

It seems prudent to give some structure to the widening B-R communication. There is a need to define what "the B-R balance of a treatment remaining positive" means, and what to do when that balance is not clear. Safety signals, and sometimes apparent benefits, continually emerge as data accumulates from growing numbers of increasingly diverse patients who, in the case of chronic therapies, use the products for longer periods of time. The term *signal* refers to events of concern that may or may not be related to the product, which require further investigation to confirm or invalidate a relationship. Clear confirmation or invalidation can be elusive, while the data continue to change as the patient population continues to diversify. It is necessary to talk about findings in a responsible manner. What is best to do? From a communications perspective, beyond improving techniques in the ways that we present benefit and risk, it will be important to agree upon the appropriate goals and venues of B-R communication. Is it helpful to merely provide information and let audiences decide what to do with it? Or is it better to make recommendations? Do the answers to these questions themselves depend upon the audience and the type of question at hand? Moving forward, we can and should continue to evolve more deliberate framing for B-R communications. It will be important to include multiple perspectives in defining what information should be communicated to support different types of decisions, and in what venues it should be made available. The process is unlikely to be easy or straightforward, but there exists tremendous opportunity and momentum as public interest has grown and stakeholders on many fronts have found common interest both in developing their understanding of other points of view and in learning to communicate across increasingly porous boundaries. We have entered an ideal time to gain unprecedented insight into assessing and understanding B-R and using that understanding to create a healthier future.

References

1. Eichler, H., Abadie, E., Baker, M., and Rasi, G. 2012. Fifty years after thalidomide; what role for drug regulators? *British Journal of Clinical Pharmacology* 74: 731–733.
2. Conan, N., Host, *Talk of the Nation*. 2013. Interview with David France, filmmaker of "How to Survive a Plague." February 12, 2013. http://m.npr.org/news/Arts+%26+Life/171406291, http://www.npr.org/2013/02/12/171406291/harrowing-stories-on-how-to-survive-a-plague.
3. Breast Cancer in Men. American Cancer Society. http://www.cancer.org/cancer/breastcancerinmen/detailedguide/breast-cancer-in-men-detection.
4. Baum, M. 2009. Breast cancer screening peril: Negative consequences of the breast screening programme. Open letter. *Times*, February 19, 2009.
5. Department of Health in association with NHS Cancer Screening Programmes, with advice and support from the Cancer Research UK Primary Care Education Research Group. 2009. Breast screening: The facts. http://www.cancerscreening.nhs.uk/breastscreen/publications/nhsbsp-the-facts-english-2009.pdf.
6. Gøtzsche, P., Hartling, O., Nielsen, M., Brodersen, J., and Jørgensen, K. 2009. Breast screening: The facts—or maybe not. *BMJ* 338: b86.
7. Gøtzsche, P., Hartling, O., Nielsen, M., and Brodersen, J. 2008. Screening for breast cancer with mammography. www.cochrane.dk.
8. Brodersen, J. 2006. Measuring psychosocial consequences of false-positive screening results—Breast cancer as an example. PhD thesis, University of Copenhagen. http://curis.ku.dk/ws/fbspretrieve/5869160/John_Brodersen_phd_afhandling_2007.
9. Brodersen, J., Thorsen, H., and Cockburn, J. 2004. The adequacy of measurement of short and long-term consequences of false-positive screening mammography. *Journal of Medical Screening* 11: 39–44.
10. Rabin, R. 2009. Benefits of mammogram under debate in Britain. *New York Times*, March 30, 2009. http://www.nytimes.com/2009/03/31/health/31mamm.html?_r=0.
11. Metcalf, M. 2009. Analyzing benefits and risks in medicine, to whom and for whom? Presented at Proceedings of the 53rd Annual Meeting of the International Society for the Systems Sciences, July 12–17, 2009. http://journals.isss.org/index.php/proceedings53rd/article/viewFile/1238/437.
12. Rabin, R. 2009. New guidelines on breast cancer draw opposition. *New York Times*, November 16, 2009.
13. Kahneman, D. 2011. *Thinking, fast and slow*. New York: Farrar, Straus and Giroux.
14. Bleyer, A., and Welch, H. 2012. Effect of three decades of screening mammography on breast-cancer incidence. *New England Journal of Medicine* 367: 1998–2005.

15. Njai, R., Siegel P., Miller, J., and Liao Y. 2011. Misclassification of survey responses and black-white disparity in mammography use, Behavioral Risk Factor Surveillance System, 1995–2006. *Preventing Chronic Disease* 8: A59.
16. Rauscher, G., Johnson, T., Cho, Y., and Walk, J. 2008. Accuracy of self-reported cancer-screening histories: A meta-analysis. *Cancer Epidemiology, Biomarkers and Prevention* 17: 748–757.
17. European Medicines Agency. Activities of patients and consumers. http://www.ema.europa.eu/ema/index.jsp?curl=pages/partners_and_networks/general/general_content_000317.jsp&mid=WC0b01ac058003500c.
18. European Patients' Academy on Therapeutic Innovation. http://www.patient-sacademy.eu/index.php/en/.
19. Patient Representatives to FDA Advisory Committees. Frequently asked questions and answers. http://www.fda.gov/ForConsumers/ByAudience/ForPatientAdvocates/PatientInvolvement/ucm123861.htm.
20. Enhancing benefit-risk assessment in regulatory decision-making. http://www.fda.gov/ForIndustry/UserFees/PrescriptionDrugUserFee/ucm326192.htm.
21. Workshop: The patient voice in clinical development: Can patients contribute to the benefit-risk assessment of new medicines? 2013. http://cirsci.org/sites/default/files/13-14%20March.pdf.
22. Temple, R. 2012. Integrating patient preferences into benefit-risk analyses. CV Safety and State of the Art Development Issues. FDA presentation, April 17, 2012.
23. Stereotactic Breast Biopsy. About.com. http://breastcancer.about.com/od/breastbiopsy/p/stereotactic.htm.

Section III

Regulatory Review
and Policy

Introduction

The handle on the door to the Regulatory Review and Policy gallery is loose, as though it has been pulled many times in many different directions. The gallery beyond the door is a round chamber. The encircling wall is unadorned and unblemished, lit coolly from below by artificial light that skitters up its smooth, white surface. In the center of the room is a large, round table, slowly rotating in a clockwise manner. There is a burnished maple rail around the table to keep viewers at a respectable distance. We are encouraged to look, but not touch.

As we approach the rail, the table tilts slightly toward us, and we can see the painting that lies flat upon it. It is an abstract piece: a startlingly detailed canvas of small color spatters. When we lean against the rail for a closer view, we hear a faint popping noise and a single drop of blue paint falls from the darkness above to yield its pigment to the painting.

7

Policy Considerations and Strategic Issues Regarding Benefit-Risk

Timothy Franson and Philip Bonforte

CONTENTS

7.1 Introduction

The World Health Organization refers to *public health* as "all organized measures (whether public or private) to prevent disease, promote health, and prolong life among the population as a whole."[*] While strengthening public health has become an almost universal objective of modern government, means used to achieve this objective vary depending on circumstance.

To scientific observers, policy deliberations involving public health might appear to be the antithesis of thoughtful clinical determinations. Policy may be made absent controlled studies or without projected impact assessments,

---

[*] http://www.who.int/trade/glossary/story076/en/.

and may often be driven by crisis situations (influenced by media attention and the forces of subjective interests). Scientific information is nevertheless valuable in policy development, although its interpretation can be profoundly influenced by context, and the perception of a policy maker can often modify or outweigh data.

Understanding policy in the context of public health, specifically drug regulation, requires understanding historical precedents and existing societal concerns. It is also critical to appreciate the framework in which policy deliberations currently take place and what consideration nonscientific policy makers may give to conflicting scientific information. These conceptual factors are well characterized in matters of benefit-risk (B-R) assessments.

The preceding chapters provide extensive details regarding definitional and operational elements of B-R for purposes of assessment and interpretation. This chapter will cover a concise review of policy factors and policy development, in addition to several aspects of B-R that may not be familiar to many healthcare practitioners. The scope of discussion is narrowly focused on Food and Drug Administration (FDA) regulation and corresponding factors within the United States. While many commonalities exist across geographic borders, it is important to understand that public health priorities, policy development processes, and similar governance mechanisms may differ across countries.

7.2 Defining Policy

The term *policy* is best defined as a principle that guides subsequent decisions to promote predictable, rational outcomes. Any given policy can be easily understood as a statement of intent from an authoritative body (e.g., federal government) that deliberately fails to stipulate operational details. Implementation of policy is carried out through legislative reform, regulation, and discretionary processes. In matters involving drug development, policy (i.e., FDA policy) is implemented via premarket approval and post-market oversight. These activities are often interlinked and result in descending tiers of implementation, each with increasing detail for interpretation and enforcement (Figure 7.1).

Policy implementation does not exist in a vacuum. Implementation efforts may be directed at subjective as well as objective decision elements. Furthermore, interpretive precedents, which tend to be more legalistic than clinical, are layered atop legislation, regulation, and like considerations.

When it comes to day-to-day activities, regulatory bodies are afforded discretion regarding enforcement and interpretation of policy. The FDA publishes internal operations guidelines (so-called *Manuals of Policies and Procedures* (MAPPs)) in order to maximize transparency and consistency in its routine review and related practices.

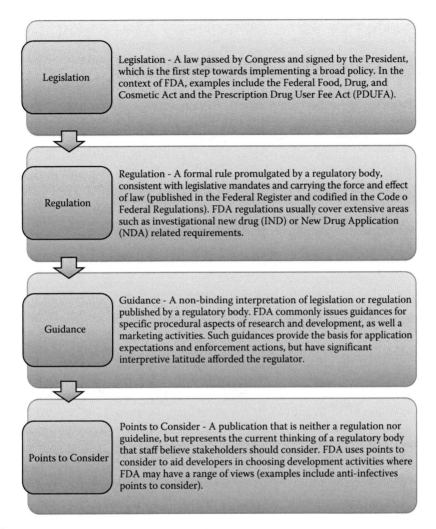

FIGURE 7.1
Descending tiers of policy implementation.

7.3 Benefit-Risk in the U.S. Regulatory Context*

In important regulatory decisions, such as for new drug approvals and post-market assessments, the concept of B-R is the fundamental yardstick

---

* While not the focus of this chapter, it is important to note that the effort to standardize B-R processes across major regulatory bodies via the International Conference on Harmonization (ICH) is an ongoing process. Without some consistency across geographies, developers will be challenged to develop efficient procedures for new compound applications that have global impact.

for deliberations. Among other requirements, a drug cannot be brought to market unless the FDA finds that there is "substantial evidence" establishing the drug as both safe and effective. The general standard for establishing efficacy has been long understood: at least two adequate, well-controlled studies. Establishing safety remains a more amorphous proposition.

Whether or not a drug is deemed safe relies on a determination that the benefits of the drug appear to outweigh its risks under the conditions of use in the proposed labeling. Critics have long viewed the FDA's benefit-risk determination as unstructured and lacking transparency. In 2007, the Institute of Medicine noted that in both the pre-approval and post-market setting, B-R assessments require a systematic process. This viewpoint exists because B-R assessments by nature are typically qualitative and relatively subjective, despite being grounded in quantitative data.

As part of the most recent Prescription Drug User Fee Act (PDUFA) reauthorization process, the Pharmaceutical Research and Manufacturers of America (PhRMA) and the Biotechnology Industry Organization (BIO) negotiated with the FDA for a more structured B-R assessment framework to improve the transparency and consistency of regulatory decision making. This negotiation resulted in the first amendment to the Food, Drug, and Cosmetic Act that expressly addresses B-R. Specifically, Section 905 of the Food and Drug Administration Safety and Innovation Act of 2012 (FDASIA), inclusive of the fifth iteration of PDUFA, requires the FDA to implement a structured B-R assessment framework in new drug approval processes.

While this is a tremendous step forward, there is ambiguity over implementation: FDASIA provides no specific mandate for regulations, guidances, or related procedures. B-R assessments will therefore remain a foundation for drug approval and post-market oversight, but the processes for standardization remain unclear, and significant discretion is left to the FDA.

The FDA has demonstrated progress in structuring B-R assessments. The agency recently developed, and is promoting among stakeholders, a conceptual 5×3 B-R assessment framework (Figures 7.2 and 7.3).

7.4 The Policy Development Cycle

Under ideal conditions, a policy is developed and implemented as part of a sequential (and ultimately cyclical) eight-stage process (Figure 7.4).

The first stage requires identifying an issue. This encompasses not only the recognition of a problem, but also an understanding of its causes and effects. One's ability to identify a discrete issue and understand the environment in which it exists can substantially impede or expedite the progress of a policy throughout the cycle.

Decision Factor	Evidence and Uncertainties	Conclusions and Reasons
Analysis of Condition	Summary of evidence:	Conclusions (implications for decision):
Unmet Medical Need	Summary of evidence:	Conclusions (implications for decision):
Benefit	Summary of evidence:	Conclusions (implications for decision):
Risk	Summary of evidence:	Conclusions (implications for decision):
Risk management	Summary of evidence:	Conclusions (implications for decision):
Benefit-Risk Summary and Assessment		

FIGURE 7.2
FDA's conceptual benefit-risk framework.

B-R Framework Designed to "Tell the Story" of the Regulatory Decision

- What is the problem?
  - *Analysis of the condition*
- What other potential interventions exist?
  - *Unmet medical need*
- What is the benefit of he proposed intervention?
  - *Benefit*
- What am I worried about?
  - *Risk*
- What can I do to mitigate/monitor those concerns?
  - *Risk management*

FIGURE 7.3
FDA's description of its conceptual benefit-risk framework.

The second stage is issue analysis. This part of the cycle delves a layer deeper in examining the issue at hand, focusing on whether a policy reform is feasible, and if so, to what extent. Determinations are made in this stage regarding the public's awareness of a particular issue, stakeholders who will need to be engaged, the means available to accomplish a solution, and potential barriers to success. The area of interest generally determines what type of analysis is conducted (e.g., objective data review or subjective judgment).

Once full analysis is conducted, a particular policy proposal can be developed based on the scope and complexity of the assessed needs. Stakeholder engagement is then initiated to "pressure test" and refine a proposed concept. Many stages of the policy cycle are tenuous, but perhaps no stage is more "make or break" than when stakeholders are formally engaged. Stakeholders frequently have varying, often conflicting, views that make it

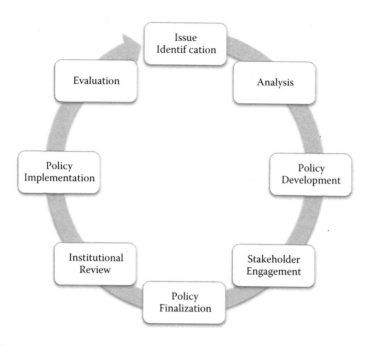

FIGURE 7.4
The eight stages of the policy development cycle.

a challenge to achieve alignment. For this reason, it is important to understand that while stakeholder engagement is essential to developing effective policy, unanimous support is often not feasible, and culminates as an explicit policy proposal.

As a potential policy moves through the cycle and gains momentum as a viable reform, it must pass through a level of bureaucracy known as institutional review. In this review process, the relevant authoritative body examines the policy proposal from a broader perspective to determine if and how it fits within established constructs. Policy development starts off as a relatively streamlined process, beginning as a concept in the mind of an individual or group. Inevitably, however, it must be approved by a hierarchical review to ensure its alignment with institutional goals and its appropriate implementation.

Policy is operationalized through implementation. To have long-term success (i.e., sustainability), a policy must be translated into concrete actions to promote its acceptance, with clarity around its scope, effects, and purpose. There are numerous ways in which a policy can be implemented. Methods include direct cut-over, where an old policy is ended at the same time a new one is enacted; parallel running, where a new policy is put into effect simultaneously with an old policy to validate its utility before a transition takes place; phased, where a policy is changed incrementally over time; and pilots, where a policy is tested before being fully implemented. The value of one approach over the others depends on both circumstance and urgency.

The final stage, evaluation, is typically ongoing and inevitably leads back to identification of new issues. Ideally, metrics for success of a policy initiative are established at the time of initial policy approval. Every component of the implemented policy is ideally reviewed for effectiveness. If a policy is not achieving its intended result on any level, evaluative findings can be used in new issue identification or in targeting implementation barriers. The main distinction between evaluation and issue identification is that the latter, by nature, is more subjective (as in whether an issue or problem is colored by social, political, or cultural influences).

7.5 Foundational Concepts

There are four key concepts that must be considered in any policy development relating to B-R: (1) accountability, (2) perception, (3) relativity, and (4) theory. Together, these concepts may be viewed as creating foundational pillars upon which a sustainable policy can be built.

7.5.1 Accountability

One might easily assume that FDA policies are developed to encompass both enabling and controlling elements. However, agency reforms are often more skewed to restrictive designs, with a focus on achieving increased oversight and enforcement. It is rare to see policy decisions that rescind prior controls, as the FDA is constrained by accountability to higher government authorities, and to the public. The FDA is a subsidiary of the Department of Health and Human Services (HHS), overseen by the Secretary of HHS, a presidential cabinet-level appointee.

Accountability is a foundational element of government, and the regulatory bodies that exist within. Society has an expectation that a regulator is serving its best interest. When such an expectation is not met, a regulator has to reclaim public trust or face substantial scrutiny (that eventually can lead to replacement of an individual or dissolution of a structure). In a similar vein, society is also skeptical of government. Change generally is not welcomed unless needed. To restate the old adage, "if it's not broke, don't fix it."

The FDA also has the difficult task of navigating public scrutiny, inclusive of the media. Historical precedent demonstrates that policy reforms are typically driven by either crises or data (Table 7.1).* Due to the high profile of threats to public health, crises are the more prevalent drivers of reform.

---

* Additionally, benefit-risk policy is rarely developed in a standard timeline. Crises can easily result in rapid legislative action, while new data may require years to permeate into mainstream policy decisions.

TABLE 7.1

Examples of FDA Policy Precedents

Trigger	Reform
Crises	
Public awareness increases regarding negligent food preparation procedures and incidences of drug addiction from patented medicine.	**Pure Food and Drug Act of 1906:** Imposed labeling requirements on drugs (and foods), laying the foundation for modern FDA oversight.
A legally marketed toxic elixir, sulfanilamide, kills 107 people, including many children.	**Federal Food, Drug, and Cosmetic Act of 1938:** Authorized FDA to, among other things, demand evidence of safety for new drugs, issue standards for food, and conduct facility inspections.
Thousands of children are born with birth defects as a result of their mothers taking thalidomide during pregnancy (for morning sickness).	**Kefauver Harris Amendments:** Required NDAs to provide evidence that a new drug is both safe and effective (the Food, Drug, and Cosmetic Act regulated only drug safety).
HIV-AIDS epidemic results in public outcry for speeding innovative therapies to market.	**Accelerated approval regulation:** Allowed for earlier approval of drugs that treat serious diseases and fill an unmet medical need based on a surrogate marker.
High-profile drug withdrawals, such as Vioxx (rofecoxib) and Avandia (rosiglitazone), prompt public scrutiny of FDA oversight.	**Food and Drug Administration Amendments Act of 2007:** Empowered the FDA with a variety of new drug safety authorities, including Risk Evaluation and Mitigation Strategies (REMSs) and the Sentinel Initiative.
Possible connection between selective serotonin reuptake inhibitors (SSRIs) and suicidal behavior in adolescents is uncovered.	**Labeling update:** The FDA proposed that makers of all antidepressant medications update their products' labeling to include warnings about increased suicidality in adolescents.
Data	
Approval rates evidence that the U.S. "lags" behind Europe in approving new drugs.	**Prescription Drug User Fee Act:** Provided a mechanism for expanding FDA resources via fees to be paid by pharmaceutical companies seeking the approval of new drugs.
Despite breakthroughs in biomedical science, the United States sees a decline in the number of innovative medical products that successfully navigate FDA approval processes.	**FDA innovation initiatives:** Established a national strategy for transforming the way FDA-regulated products are developed, evaluated, and manufactured.

How the FDA responds to a crisis or data can be difficult to generalize. Ideally, the agency waits to react until a major problem unfolds or it has evaluative findings. This reaction can be broadly characterized as first correcting the problem at hand and then attempting to prevent its reoccurrence. This is similar to the principle of corrective and preventive actions that are part of quality management systems in most manufacturing operations.

7.5.2 Perception

Although crises may trigger public calls for reform, it is the perception of policy makers that more often shapes outcomes. A noteworthy difference exists between how clinical professionals and policy makers perceive crises in the context of biological and pharmaceutical products. A clinical professional's perception is based on accumulated knowledge of, and experience in, healthcare; however, a policy maker's perception is often based on a collection of knowledge and experience outside of healthcare, such as finance, transportation, and national security.*

Full appreciation of how perception affects decision making requires profiling the disconnect that may exist between policy makers' and clinical professionals' views of the same issue. The concept of safety, a foundational component of B-R, is a prime illustration of such a disconnect.

Relying on knowledge and experiences outside of healthcare, policy makers often perceive achieving safety as an absolute outcome, a state of being completely protected from a recognized harm. The Delaney Clause of the Federal Food, Drug, and Cosmetic Act serves as a prime example of this perception. Enacted in 1958 after one of the first food scares in modern history based on a chemical additive, the Delaney Clause bans any food additive found to cause cancer in man or animals. This is essentially an attempt to achieve absolute protection from a recognized harm (i.e., carcinogens). For broader examples outside of the FDA context, one need only look at recent policies developed for school and work environments. The normal frame of reference for matters such as work and school safety programs is to achieve complete freedom from harm, and that such goals are achievable. Indeed, many dictionaries define *safety* as the reciprocal of risk, or freedom from risk. Such a perception of safety does not easily translate to drugs.

Clinical professionals understand that health exists on a continuum where one's desire to be protected from harm is not absolute and varies depending on circumstance. Further, medical professionals, drug developers, and the FDA recognize that every drug has risks, and thus safety is a concept distinct from other societal situations. For these reasons, they perceive safety as a malleable concept, the control of recognized harm when juxtaposed with a potential benefit, for the purpose of achieving an acceptable level of risk for a given patient.

---

* In all such settings, risk mitigation policies generally follow crises (one need only look at the financial crises of 2008 or Three Mile Island for relevant examples).

The result of this disconnect is that when policy makers respond to a crisis involving a pharmaceutical product, promulgated policies often fail to be appropriately tailored to the unique characteristics of healthcare.

7.5.3 Relativity

One of the complexities around how clinical professionals interpret safety is that "acceptable risk" is a relative outcome. What constitutes an acceptable risk can easily change from stakeholder to stakeholder, whereas absolute protection from a recognized harm is a clearly defined goal that leaves little room for ambiguity (but is rarely achievable in healthcare). For example, a patient with a terminal illness and limited options for treatment is likely to find a high-risk drug more acceptable than a healthy consumer who does not require medical intervention on a regular basis.* Additionally, a patient who measures acceptable risk in the context of therapeutic response is distinct from a payer who may look at risk in the context of cost. A key distinction is that policies usually focus on population impact; conversely, practitioners desire a breadth of information as a frame of reference for individual patient decisions.

As previously discussed, policy makers traditionally have accumulated experience and knowledge from settings outside healthcare. While acceptable risk is ingrained in clinical judgment, the relativity that exists in defining *acceptable* creates significant burden for those unfamiliar with healthcare to embrace policies that stem from a clinical view of safety. Any success in developing policies around acceptable risk will inevitably require utilization of broad societal reference points and cross-learning from disciplines beyond healthcare in order to mitigate the burden of (and hesitation toward) embracing relative outcomes.

7.5.4 Theory

Development of B-R policy encompasses a number of inherent theoretical questions. Comprehensively addressing these challenging queries is beyond this chapter's scope, but warrants brief mention, as they will likely affect future policy development. It is important to note that many of these questions are driven by how varying stakeholders perceive benefit and risk.

- *What is the difference in assessing routine risk versus life-threatening risk?*[†]
  Every activity, process, or product encompasses risk. While event

---

* Acceptable risk among patients with the same disease may also vary depending on personal circumstance. An individual diagnosed with leukemia at the age of 20 may be more likely to desire a treatment that ensures his longevity regardless of risks to quality of life; however, an individual with the same diagnosis at the age of 80 may place a higher priority on maintaining current quality of life.

[†] http://web.iitd.ac.in/~arunku/files/CEL899_Y12/Understanding%20life-threatening%20 risks_Keeney.pdf.

1. Existence of Risk	2. Analysis of Risk
Life is not and cannot be free of risk	*The existence of risk in life can never be eliminated, only transferred*

3. Evaluation of Risk	4. Choice of Alternatives
Risk can be defined by many circumstances (economic, political, social, etc.)	*When benefits are considered, alternatives with greater risk may be preferred to alternatives with less risk*

FIGURE 7.5
Four factors for making the distinction between routine and life-threatening risk.

frequency is a concern, it is the severity of such risk that informs development of policy and validates outcomes. The difference between routine and life-threatening risk is the probability of death. There are gray areas in the spectrum of routine and life-threatening risk; without a unified method of appraisal, classifications are largely subjective, and thus provide an unstable foundation for policy development. Four factors have been used to characterize the distinction between routine and life-threatening risk (Figure 7.5). These factors all warrant careful consideration when changes in B-R policies are entertained.

- *What role do complex systems play in managing B-R?* A complex system is a network of varying component parts (processes, people, etc.) that interact to give rise to unpredictable behavior. Understanding complex systems and their role in causing and mitigating risk is important to consider in B-R policy development as healthcare infrastructures span multiple practice specialties, practice guidelines, and potentially conflicting incentives. This is an especially sensitive dynamic when assessing an isolated safety issue. For example, B-R decisions in the setting of an intensive care unit are inherently more complicated than traditional outpatient settings (if for no other reason than log scale differences in comorbidities, concomitant medicines, and intercurrent procedures in intensive care that lead to higher likelihood of drug-drug, drug-disease, and drug-device interactions).

- *How do crises lead to tipping points in policy?*[*] There are multiple examples of seemingly discrete events leading to broader policy development. Whether an event (or events) results in policy development can depend on whether it is magnified over time to become a mainstream issue. Amplifiers (such as the press) and visible spokespersons may be very persuasive with policy makers and can be pivotal in creating these tipping points. The Pure Food and Drug Act of

---

[*] Gladwell, Malcolm, *The Tipping Point: How Little Things Can Make a Big Difference* (Boston: Little, Brown, and Company, 2000).

1906, one of the foundations of FDA authority, was precipitated by "muckraking" journalists exposing issues in the food and pharmaceutical industry.

- *How does the precautionary principle influence perception?*[*] Primarily a European concept, the precautionary principle postulates that a new technology should be integrated into settings (medical or otherwise) only if there is proof that it will not cause unacceptable risk. In the extreme, if one is forced to "disprove a negative" in the context of drug development, clinical trial sizes and approval timelines could become too burdensome (and ultimately impractical). An alternative concept would be to characterize ranges of acceptable risk for a given setting, then assess whether a new technology fits within that range. This approach seems well suited for development of new drugs for rare diseases, where no therapies may be available and patients may be willing to accept appreciable risks for the hope of some benefit.

- *What are the barriers to accruing evidence in unique situations (such as rare disease)?*[†] Understanding unique health circumstances and accounting for their consequences are necessary components of developing appropriate, targeted B-R policy. For most rare diseases, there are few baseline epidemiologic studies of natural history, no comparators against which to measure the effect of treatment, and much smaller pools of patients for trial enrollment. Thus, when comparing prevalent and rare diseases, the ability to achieve a similar burden of proof for benefit and risk factors may not be possible.

- *Should benefit-risk be viewed as an integrated concept?* The concept of benefit-risk evokes a "bipolar" response from numerous authorities. Some experts believe that benefit and risk are "opposing forces" that must be separately assessed, while others opine that some degree of interwoven logic is essential (e.g., at the patient level, decisions relating to benefit and risk must be weighed concurrently). Such a clinical dilemma is beyond the scope of this chapter; however, it is important to have an awareness of the conceptual discord to ensure that developed policies do not limit considerations or restrict future insights.

7.6 Prevailing Issues

Prior to 1992, the FDA's drug regulatory processes were often described as slow and inadequately staffed. The AIDS epidemic, coupled with a

---

[*] http://www.sciencedirect.com/science/article/pii/S0957582003710647; https://www.wbginvestmentclimate.org/uploads/lofstedt.pdf.
[†] http://www.rarediseases.org/docs/policy/NORDstudyofFDAapprovaloforphandrugs.pdf.

perception that Europe had greater access to medical innovation, substantially changed the regulatory environment for drugs in the United States. Through enactment of PDUFA in 1992, and its subsequent reauthorizations in 1997 and 2002, the FDA implicitly embraced balanced consideration of B-R in programs such as accelerated drug approval, priority review, and fast-track designation. For over a decade, innovative drugs reached those in need and did so at a faster rate than in other countries.

Progress slowed after the Food and Drug Administration Amendments Act of 2007 (FDAAA), which included the fourth iteration of PDUFA. Public anxiety from safety issues with high-profile drugs, such as Vioxx (rofecoxib) and Avandia (rosiglitazone), led to the withdrawal or restricted use of such drugs and congressional scrutiny over the rigor exercised by the FDA to protect public health. FDAAA resulted in a significant focus on risk management, especially with regard to intensified post-market interventions. Among other reforms, Congress established an FDA authority to require a Risk Evaluation and Mitigation Strategy (REMS) for any new drug application (NDA), abbreviated new drug application (ANDA), or biologic license application (BLA) at any stage of the product life cycle. The FDA also reacted with a series of discretionary actions that represented a shift toward risk management, including the Safety First Initiative and the Center for Drug Evaluation and Research's (CDER) creation of "super offices" focused on safety and enforcement.

Today, the FDA's B-R policies continue to evolve. The previously mentioned conceptual B-R framework represents a significant step forward. Nevertheless, unresolved issues exist for structural and interpretive aspects of B-R governance and remain a challenge in B-R policy development.

7.6.1 Risk Aversion

It can be argued that the burden of evidence needed for the agency to act on a concern over a new risk is quite different from its standard for approving a claim for a new benefit. The FDA typically requires two adequate and well-controlled studies to even apply for a new labeling claim, which may require years and appreciable investment. This standard can be juxtaposed against the fact that the FDA may require a post-approval labeling change or restrict usage based only on a single observation of a serious adverse event, which can occur in a matter of days.[*]

Risk aversion at the agency level can also be surmised from post-approval requirements. The majority of post-approval studies required by regulators are based on concern over a drug's risk. Conversely, few post-approval studies are mandated by regulators to assess a drug's benefit. As a result of this disparity, it is almost inevitable that new post-approval findings will provide

---

[*] A notable illustration of dramatic regulatory interventions is that of new hepatotoxicity findings (either pre- or post-market) that suggest the potential for severe adverse events in patients may trigger prompt regulatory restrictive actions.

a preponderance of new adverse event data, and thereby appear to nega-tively skew the balance of a drug's B-R profile.

This disproportional trend is rooted in policies stemming not from FDAAA, but the earlier Public Health Security and Bioterrorism Preparedness and Response Act of 2002 (PDUFA III). PDUFA III which included the third itera-tion of PDUFA required post-approval risk management reporting for drugs, but no such parallel efforts for gathering data on benefit. Risk assessment and management policies have since grown, inherently tilting away from a balanced weighing of B-R, and thus affecting agency action trends. As an example, many safety communications from the FDA present only new adverse event data without fair balance in a discussion of B-R.

7.6.2 Patient Perspectives

The FDA and industry have inherent biases regarding drug approval pro-cesses: the FDA worries foremost about risks; industry focuses on benefit profiles. To an extent, these biases are natural. However, merely having two ends of a spectrum emphasized does not necessarily lead to acceptable bal-ance. Patient perspectives are a key missing element in achieving more ratio-nal, unbiased policy outcomes.

No party is better able to assess a drug's acceptable risk than those living with the illness the drug is designed to address. Patients can be a respected voice of reason in decision-making processes, tempering industry and regu-latory predispositions to ensure progress. Following the HIV epidemic of the 1980s, patient perspectives were included in the drug approval process for antiviral medicines; this contributed to timely approval of the first protease inhibitor in 1995, and in 97 days, AIDS went from a seemingly fatal disease to a chronic illness.

The FDA has recently begun to incorporate patient views in drug devel-opment, via patient-focused drug development programs, patient-reported outcome (PRO) tools, and the recently announced FDA-sponsored Patient-Focused Drug Development Initiative. However, these efforts are in very early stages, and their impact on B-R remains to be seen. Patient-centric poli-cies will be closely scrutinized as a bellwether for future policy initiatives.

7.6.3 Quantitative Assessments

The efficiency of a regulatory process hinges on one's ability to understand and navigate through it. In order to foster consistency and predictability in such assessments, stakeholders need a decision-making process based on agreed definitions, criteria, and methods for quantifying data.

Defining a drug's benefits and risks encompasses a wide spectrum of con-siderations. Weighting these criteria's relative importance is naturally subjec-tive, but inconsistent outcomes cause chilling effects in drug development, especially when dramatic actions for specific compounds are viewed by industry as onerous precedents that can discourage development programs. A quantitative assessment framework allows value judgments to be objectively assessed, alongside efficacy and safety data, in a well-articulated and transparent process. Developing such a framework is challenging, but not impossible. The European Medicines Agency (EMA) has been vetting quantitative methodologies since 2008 as part of its Benefit-Risk Methodology Project.

Industry and global regulators are increasingly exploring quantitative analytics in an attempt to move away from more empiric, conceptual approaches to B-R. If the United States is to remain a competitive regulatory environment, the FDA will likely need to become more engaged in shaping this field of study.

7.6.4 Harmonization

When comparing the FDA B-R policy to that of the EMA, there are multiple policy areas exhibiting a clear lack of concordance in regulatory decisions. Concern exists over the differing B-R decisions among the EMA and FDA, and the confusion they create among practitioners and patients about which regulator is "right." This situation is well depicted in a contrasting decision on a new obesity compound: the FDA approved Qsymia (phentermine and topiramate) and stated that those deliberations could be a model for B-R decisions, while the EMA declined to approve this drug on the grounds that its assessment of risks exceeded benefits. A second example of drastically divergent decisions relates to salmon calcitonin-containing drugs. The EMA has withdrawn salmon calcitonin-containing drugs on oncogenesis concerns, while the FDA has taken no similar actions (nor has the FDA relabeled these products to note heightened risk).*

Whether the aforementioned decisions were influenced by science, advocacy, media, or other forces will remain a topic for discussion; the existence of differences in B-R assessments does not necessarily derive from policy disparities. Nevertheless, the divergence clearly illustrates that reasonable scientists may differ on assessing B-R for the same drug. Policy can play a key role in influencing consistent approaches to data analysis (itself a form of harmonization).

7.7 Next Steps

Advancing B-R policy development consists of short- and long-term processes requiring both situational awareness and well-timed action. A policy

---

* The FDA convened an advisory committee in March 2013 which recommended salmon calcitonin no longer be marketed in the US. As of August 2013, the FDA had not taken final action in this matter.

strategy must embrace opportunities at every level (globally, regionally, and among segmented stakeholders). Five broad needs should be considered for future B-R policy development:

1. **A common vocabulary:** Having clearly defined terminology helps combat diverging perceptions, ultimately making it easier to align stakeholders and promote buy-in.

2. **A common assessment framework:** Consistent B-R decisions facilitate regulatory certainty and enable more transparent decision processes.

3. **A disciplined assessment infrastructure:** Appropriately responding to crises requires a measured assessment of circumstances to ensure reactions are not overly broad.

4. **A life cycle approach:** The B-R profile of a drug can change over time; mechanisms are needed to facilitate shared learning about a drug's benefits and risks as these factors evolve.*

5. **Balanced safety communications:** Making an informed decision about a drug's safety requires being informed about not just new risks (i.e., adverse events), but also corresponding updates about benefits.

In addition to existing needs, outstanding questions exist. The use of quantitative analytics raises concern about how B-R considerations will be weighted (and how such weighting will be communicated in labeling). There is also uncertainty about how to achieve more balance in gathering and sharing information about post-approval data.† Policy development will be significantly shaped by whether, and how, these questions are answered.

7.8 Summary Points

Generally speaking, crisis events are not representative of most drugs' B-R profile. As a result, many feel that the reactive nature of B-R policy development, when based on episodic product safety crises, often results in overly broad restrictions and a public perception of unsafe drugs in the market. A more disciplined policy assessment infrastructure would ideally channel

---

* As an example, for the IND phase of studies, the use of biomarkers and disease models can be advanced from early-phase trial work to applicability in patient care settings (this may be a formidable challenge for those laboring in the field of translational medicine).

† This consideration is vital for labeling purposes. As previously noted, considerable post-marketing data collection requirements for all compounds ensure that voluminous data relating to risk will be accrued by regulators over time. The "tipping point" for a new safety observation in labeling may be a single serious unexpected adverse event, whereas a new benefit claim requires a substantially higher burden of evidence.

Perspective
Stakeholders commonly view benefi -risk issues differently; understanding such divergence is necessary to ensure all existing considerations receive appropriate attention.

Product profile
Special consideration should be given to benefi -risk in the context of a drug targeting an unmet patient need, as in rare diseases, resistant microorganisms and cancers refractory to conventional treatment.

Population
How a particular crisis impacts the population (at-large and in terms of a particular patient group) should be the starting point for assessing the appropriateness of a particular policy response, including potential indirect or unintended effects.

Publicity
The influe e of media, consumer advocates, patient advocates and the like is highly dependent on how the experts at FDA and sponsors communicate (both to each other and externally).

FIGURE 7.6
Core P considerations for benefit-risk policy development.

urgent matters in balanced fashion and ensure periodic reassessments for efficacy of approved policies.

In constructing ways forward for such an improved infrastructure, it may be possible to encourage more thoughtful consideration of B-R policy by reemphasizing core P considerations (not to be confused with P values used in statistical appraisal of clinical benefit-risk data). The principle factors for B-R policy, which begin with the letter P, are perspective, product profile, population, and publicity (Figure 7.6). All such P considerations deserve careful attention before pursuing policy changes.

It may be challenging to distill points made in earlier sections to a summary set of principles. However, it is clear that the circumstances of balancing public health with unique needs of varying stakeholders (patients, consumers, etc.) demands more disciplined and creative approaches to B-R policy development.

Any creative policy can be initially calibrated for discussion purposes by considering how such policy varies depending on disease type, including "severity strata," such as life-threatening diseases, progressive chronic serious diseases, diseases altering patient quality of life (not routinely fatal but constraining activities of daily living), and diseases of life preferences (as with erectile dysfunction, toenail fungus, or other lifestyle disorders that do not compromise normal life actions). By having reference points relating to B-R trade-offs, it may be easier to facilitate constructive exchanges among competing views.[*]

---

[*] Such reference points may also serve as homeostatic mechanisms to preclude politically driven policy actions without data or impact considerations.

Ultimately, patients depend on rational policy decisions to complement good data, and the evolving aspects of B-R policies are most deserving of more intensive development in the coming months and years. More thoughtful and deliberate structured approaches to policy changes should be an aspirational goal for all stakeholders in the drug development and commercialization enterprises.

Reference Sources

The authors of this chapter relied upon information from selected references that apply broadly to multiple regulatory policy topics covered in this review. Interested readers are invited to further refer to those sources of general information, which are listed below.

1. Gladwell, M. 2000. *The tipping point: How little things can make a big difference.* Boston: Little, Brown, and Company.
2. http://dij.sagepub.com/content/46/6/736.
3. http://healthland.time.com/2012/10/04/meningitis-outbreak-steroid-shots-highlight-dangers-of-compounding-pharmacies/.
4. http://healthyliving.msn.com/health-wellness/fixing-a-failing-fda-1?pageart=2.
5. http://jama.jamanetwork.com/article.aspx?articleid=1360873.
6. http://oversight.house.gov/wp-content/uploads/2012/01/4-21-11_Roth_SD_Testimony.pdf.
7. http://sphhs.gwu.edu/releases/obesitydrugmeasures.pdf.
8. http://www.biopharminternational.com/biopharm/News/FDA-Office-of-Compliance-to-Become-a-Super-Office/ArticleStandard/Article/detail/726810.
9. http://www.chi.org/uploadedFiles/Industry_at_a_glance/Competitiveness_and_Regulation_The_Future_of_America's_Biomedical_Industry.pdf.
10. http://www.compliancearchitects.com/2012/01/fda-warning-letters-increase-155-from-2010-levels/?utm_source=rss&utm_medium=rss&utm_campaign=fda-warning-letters-increase-155-from-2010-levels.
11. http://www.fda.gov/AboutFDA/CentersOffices/OfficeofMedicalProductsandTobacco/CDER/WhatWeDo/Initiatives/default.htm.
12. http://www.fda.gov/AboutFDA/ReportsManualsForms/Reports/ucm276457.htm.
13. http://www.fda.gov/AboutFDA/WhatWeDo/History/Overviews/ucm305697.htm.
14. http://www.fda.gov/downloads/AboutFDA/CentersOffices/OfficeofMedicalProductsandTobacco/CDER/UCM303351.pdf.
15. http://www.fda.gov/downloads/AboutFDA/CentersOffices/OfficeofMedicalProductsandTobacco/CDER/UCM306183.pdf.

16. http://www.fda.gov/downloads/AboutFDA/CentersOffices/OfficeofMedicalProductsandTobacco/CDER/UCM310754.pdf.
17. http://www.fda.gov/downloads/AboutFDA/CentersOffices/CDER/UCM309983.pdf.
18. http://www.fda.gov/downloads/Drugs/GuidanceComplianceRegulatoryInformation/Guidances/UCM267449.pdf.
19. http://www.fda.gov/downloads/ForIndustry/UserFees/PrescriptionDrugUserFee/UCM270412.pdf.
20. http://www.fda.gov/downloads/ForIndustry/UserFees/PrescriptionDrugUserFee/UCM279724.pdf.
21. http://www.fda.gov/downloads/ScienceResearch/SpecialTopics/RegulatoryScience/UCM268225.pdf.
22. http://www.fda.gov/ForIndustry/UserFees/PrescriptionDrugUserFee/ucm326192.htm.
23. http://www.fda.gov/NewsEvents/Newsroom/PressAnnouncements/ucm216448.htm.
24. http://www.fda.gov/NewsEvents/Speeches/ucm279777.htm.
25. http://www.fda.gov/RegulatoryInformation/Legislation/FederalFoodDrugandCosmeticActFDCAct/SignificantAmendmentstotheFDCAct/PrescriptionDrugAmendmentsof1992PrescriptionDrugUserFeeActof1992/default.htm.
26. http://www.fda.gov/ScienceResearch/SpecialTopics/CriticalPathInitiative/ucm076689.htm.
27. http://www.fda.gov/ScienceResearch/SpecialTopics/RegulatoryScience/default.htm.
28. http://www.fdli.org/resources/resources-order-box-detail-view/how-should-fda-align-benefit-risk-drug-evaluations-stipulated-by-pdufa-v-with-the-needs-of-the-end-users-patients.
29. http://www.forbes.com/sites/matthewherper/2012/12/05/new-drug-approval-rate-at-near-record-high-fda-says/.
30. http://www.gpo.gov/fdsys/pkg/PLAW-112publ144/pdf/PLAW-112publ144.pdf.
31. http://www.grassley.senate.gov/news/Article.cfm?customel_dataPageID_1502=40602.
32. http://www.idsociety.org/uploadedFiles/IDSA/News_and_Publications/IDSA_News_Releases/2012/LPAD%20FAQs.pdf.
33. http://www.medicalprogresstoday.com/fda-regulation-medical-innovation/.
34. http://www.nytimes.com/2005/03/02/politics/02fda.html.
35. http://www.phrma.org/media/releases/phrma-transitions-management-benefit-risk-action-team-framework-cirs.
36. http://www.pwc.com/us/en/health-industries/payers/challenges-faced-by-payers.jhtml.
37. http://www.rarediseases.org/docs/policy/NORDstudyofFDAapprovaloforphandrugs.pdf.
38. http://www.sciencedirect.com/science/article/pii/S0957582003710647.
39. https://www.wbginvestmentclimate.org/uploads/lofstedt.pdf.
40. http://web.iitd.ac.in/~arunku/files/CEL899_Y12/Understanding%20life-threatening%20risks_Keeney.pdf.
41. http://www.who.int/trade/glossary/story076/en/.
42. Slovic, P. 2000. *The perception of risk*. London: Earthscan.

8

Systematic Approaches to Benefit-Risk Assessment

Rebecca Noel

CONTENTS

8.1 Introduction

As highlighted throughout this book, assessing the benefit-risk (B-R) balance or profile of treatment options, be they drugs, devices, or procedures, is a prominent challenge across the spectrum of healthcare and drug development. B-R assessments have been conducted since the modernization of the drug regulatory system in the wake of the 1960s thalidomide disaster; however, it has only been within recent years that progress has been made on apposite methodologies for conducting formal assessments.[1-5] Chapter 7 provided a concise review of policy development; this chapter reviews progress made toward implementing structured and consistent approaches to B-R assessment. It highlights key developments in systematic B-R assessment.

8.2 Benefit-Risk Balance for Marketed Drugs: Evaluating Safety Signals

In 1998, the Council for International Organizations of Medical Sciences (CIOMS) Working Group IV released its report on assessing the B-R balance of marketed drugs.[4] In its report, the working group's conclusion foreshadowed the discussion that was to begin anew approximately a decade later:

> The comparative evaluation or weighing of benefits (positive effects) and risks (potential harm) of various medical options for treatment, prophylaxis, prevention or diagnosis is essential. It is done during research and development on new medical products or procedures (such as surgery), or by a regulatory authority deliberating the approval or withdrawal of a product or some intermediate action, by a physician on behalf of a patient, or by the patient. Such weighing, whether implicit or explicit, is at the heart of decision making in medicine and health care.
>
> This apparently straightforward concept is expressed through such terms as benefit to risk ratio, benefit-risk difference, benefit vs. risk, therapeutic margin, therapeutic index and others. Regrettably, in spite of common and frequent use, none of these inexact expressions has been adequately defined or is easily quantifiable with a summary statistic. Moreover, although regulators and companies routinely make decisions driven by the balance between benefits and risks, however they are measured, there are no generally agreed procedures or regulatory guidelines for conducting and acting upon benefit-risk assessment.

While the CIOMS report focused on evaluating the benefits and risks of a marketed product within the specific context of pharmacovigilance activities (i.e., at the time of a suspected drug safety concern), the working group also described its larger vision for the future state of B-R assessment in general, expressing a desire that both developers and regulators adopt a systematic approach. The CIOMS group put forward a set of key principles and practices to guide a B-R assessment. These principles and practices are generally summarized as:

- Decisions are made on behalf of an at-risk population, and from a public health, as opposed to individual patient, perspective. However, the needs and perspectives of different stakeholders should be taken into account.
- A B-R evaluation should be conducted relative to either no treatment or appropriately selected comparators. Standardized graphical representations should be used to facilitate comparisons between alternative treatments.
- A B-R assessment should be holistic, in that individual benefits or risks should not be evaluated in isolation. In the context of a

marketed product, a new safety issue should prompt a reevaluation of the entire safety profile, or at least of prominent or important
adverse drug reactions relative to other treatments.

- The basis for, rationale of decisions to be made, and processes com-
prising a B-R assessment should be transparent.[4]

8.3 The Future of Drug Safety: Promoting and Protecting the Health of the Public

The 2004 withdrawal of Vioxx (rofecoxib)[6] and other similar high-profile
safety alarms brought the U.S. Food and Drug Administration's (FDA)
efforts to ensure drug safety under considerable public scrutiny that eventually escalated to Senate hearings. The safety controversies contributed to
a public perception of a drug safety crisis. However, with the perceived crisis came the opportunity for a thorough evaluation of the U.S. drug safety
system. As a step toward restoring confidence in the agency, the FDA asked
the Institute of Medicine (IOM) for a comprehensive evaluation of the existing drug safety system with a view toward identifying areas of vulner-
ability and facets of the system that could be strengthened to improve its
ability to meet the needs of the American public. Nearly a decade after
the CIOMS report, B-R assessment was once again front and center in the
public forum.

In its 2006 report, *The Future of Drug Safety*, the IOM committee noted that
reviewers at the FDA, and the Center for Drug Evaluation and Research
(CDER) in particular, are required to weigh all available information about
a drug's benefits and risks, make decisions in the context of scientific uncertainty, and integrate information on a drug's B-R profile throughout its
life cycle. In assessing a new drug application (NDA), the IOM committee
observed that FDA clinical reviewers are required to analyze the efficacy
and safety data from clinical trials, as well as evaluate the robustness and
appropriateness of the statistical methods used by the drug sponsor. To
assist reviewers with this daunting task, the FDA and CDER adhere to a multitude of guidances, policies, and procedures that outline best practices for
reviewing and analyzing the data provided in a submission dossier. Despite
the availability of these tools and resources, the IOM committee concluded
that "in both the pre-approval and the post-marketing setting, the risk-benefit analysis that currently goes into FDA decisions appears to be ad hoc,
informal, and qualitative." Like the CIOMS, the IOM report focused on drug
safety. Nevertheless, the IOM committee envisioned a transformed assessment system and called for an overarching, life cycle-based approach to the
assessment and management of B-R.

The IOM found that a life cycle approach had not been widely embraced by the FDA, and as a result, had been implemented in a limited and fragmented manner. For the FDA to fully adopt and employ a life cycle approach, the IOM committee believed the agency would need to dedicate continued attention to a drug's B-R profile over the course of its life. This would require the agency to continually monitor new data and engage in ongoing, active reassessment of benefits and risks that might spark regulatory action. These activities would need to be made possible by enhanced pre- and post-approval regulatory authority.[7]

8.4 Benefit-Risk Assessment: The Spark Ignites

The period 2005–2007 was important for B-R assessment: regulators, governmental authorities, payers, and the public all seemed to suddenly request additional access and information on the benefits and risks of drugs. In addition to the IOM report, a similarly timed U.S. Government Accountability Office (GAO) report also questioned the FDA's post-marketing safety activities.[6,8] Drug withdrawals coupled with the GAO and IOM reports served to focus a multitude of interested parties on the future of B-R assessment, including regulators, the Pharmaceutical Research and Manufacturers of America (PhRMA), industry, think tanks, and legislators.

In Europe, the European Medicines Agency (EMA) was developing a reflection paper on B-R assessment. Prior to issuing the paper, the EMA's Committee for Medicinal Products for Human Use (CHMP) established a working group to provide recommendations to improve B-R methodology, as well as the consistency, transparency, and communication of B-R assessments.

The CHMP working group reviewed established methods of B-R assessment and considered their practical applicability. The group concluded that expert judgment was expected to remain the cornerstone of B-R evaluation for the EMA authorization. The group did not anticipate quantitative B-R assessment replacing qualitative evaluation. While the reflection paper did not define "quantitative B-R assessment," it did provide a list of examples.

The working group put forward two main recommendations. First, the CHMP should revise the B-R assessment section of its assessment report templates, incorporating a structured list of benefit and risk criteria and guidance. The proposed modifications were:

- To use a structured and mainly qualitative approach
- To be explicit about the importance of benefits and risks in the specific therapeutic context
- To describe sources of uncertainty and variability and their impact on the B-R assessment

The working group's second recommendation was that the EMA should, in consultation with other experts and EMA regulators, continue to pursue research on methodologies of B-R assessment.[9]

Outside of the EMA, the Centre for Innovation in Regulatory Science (CIRS)* engaged a consortium of regulators from four countries—Canada, Australia, Singapore, and Switzerland—to develop a framework to support the sharing of work products relevant to B-R assessment among the agencies, communicate decisions to stakeholders, and develop documentation tools for use during regulatory review. The CIRS referred to this initiative as the Consortium on Benefit-Risk Assessment (COBRA). In addition to their work with these four regulatory agencies, CIRS also acted as a forum to bring together various parties within industry, regulatory agencies, and academe to discuss how to advance the science and practice of B-R assessment. CIRS originally became active in B-R assessment via an interest in the application of multicriteria decision analysis (MCDA) models.[10] As discussed in Chapter 5, MCDA models provide a way to evaluate large, complex problems by decomposing them into smaller, more manageable pieces that can be studied using a mixture of data and judgment. These pieces can then be reassembled to provide insight to help inform decision makers.[11] As their experience and interaction with other B-R assessment initiatives grew, however, CIRS expanded its thinking and moved beyond the strict application of MCDA to embrace the broader concept of a structured or framework-based approach to B-R assessment. Under this aegis, MCDA became one of many methodologies available to complement a B-R framework.[12]

Advancements in pharmacovigilance, coupled with a spate of high-profile safety crises and a review of regulatory practices by independent agencies and stakeholders, highlighted existing gaps in knowledge and application of B-R assessment. As a result, B-R had become one of the more pressing issues in drug development and regulatory review. Ironically, a positive B-R profile had been a required standard for approval and marketability all along. For a new drug or device to gain and maintain market access in the United States, it must demonstrate compliance to manufacturing standards, an acceptable safety profile for its indication, and a proven effect on specific efficacy parameters. Atop these three criteria (often referred to as the three pillars of approval) rests the concept of B-R balance, which must be positive in order for a product to obtain marketing authorization. Similar regulatory standards exist across the world. For example, in the European Union (EU), the EMA is required to make its decisions based on a drug's overall B-R assessment (Article 26, Directive 2001/83).[13]

---

* The Centre for Innovation in Regulatory Science (CIRS) was formerly known as the Centre for Medicines Research (CMR).

8.5 The Benefit-Risk Assessment Paradox

B-R had been a fundamental regulatory consideration, and B-R balance had become the subject of much discussion; however, it still was not clear exactly how a decision maker was supposed to move from an evaluation of data to a summary B-R conclusion. What was that process? And what did it entail? These questions posed a real dilemma. The reality was that although the assessment of a drug or device's B-R profile was critical for decision making, the actual process of bringing together the holistic assessments of quality, safety, and efficacy was ill-defined, and remained ill-defined until very recently.

Publically available documents contained no references to any agreed upon or defined approach. Indeed, the historical record implied that the FDA was more influenced by reaction to external events and political and stakeholder pressure than adherence to a formal process. Lacking appropriate tools and processes to integrate, weigh, and assess data, or to consider the perspectives of various stakeholders, regulatory agencies relied primarily on expert judgment from their staffs and advisory committees. As a result, B-R assessments were poorly defined, inconsistent, unpredictable, and opaque processes that often disproportionately emphasized risks at the expense of benefits.

One can argue that the failure of pharmaceutical B-R assessment to develop a formal process to guide or complement clinical judgment can be traced to two main limitations: complexity and conflict. Early attempts to build and advance quantitative methodologies were too specialized, overly narrow, and incomprehensible to the majority of potential users. Exacerbating this was a tension between a paternalistic regulatory mindset and desire for increased transparency.

A quick survey of the literature prior to 2006 yields a host of quantitative methods developed specifically for use in pharmaceutical B-R assessment. An early approach by Tallarida incorporated the severity of side effects with the simultaneous assessment of risk and benefit.[14] The resulting scale divided severity into seven classes of intensity. The risks and benefits of a drug were subsequently expressed in terms of expectations using probabilities and scores assigned to the seven severity classes. In the case of multiple adverse reactions, the risk was calculated as the sum of their individual expectations. The difference between the expected benefit and the expected risk was calculated as benefit minus risk, or net benefit.

The concept of discounting benefit measures to account for negative or adverse events was employed by several other quantitative approaches, as was the notion of modifying measures to adjust for changes in quality of life. Gelber et al.[15] developed TWiST (time without symptoms or toxicity). TWiST presented a measure of B-R based on time without symptoms of disease and toxic effects. Quality of life adjustments were later included, resulting in Q-TWiST (quality-adjusted time without symptoms or toxicity). Both

TWiST and Q-TWiST partition overall survival into various states of health, use Kaplan-Meier methods to estimate the mean duration of those health states, assign weights to each of the states, and calculate a value for each treatment group that can be used to make comparisons across groups.

Chuang-Stein developed the global risk-benefit (GRB) methodology to quantitatively describe the risk-benefit of an intervention by comparing differences in GRB scores using an asymptotic statistical distribution.[16] The GRB procedure consists of creating composite GRB categories, partitioning patients into the categories, and making statistical comparisons of treatment group differences based on GRB scores. Though clearly useful in certain applications, the Chuang-Stein approach is limited by the clarity of demarcations between benefit and risk, and is computationally intense.

The specialized approaches described above have found limited utility as applied B-R tools. Industry and regulators were not lacking methods, but an overarching, systematic process or framework that could be used to guide and structure their B-R process. Rather than looking to quantitative methods as the answer, what they needed first was a set of guidelines and processes to direct and support transparent, systematic, higher-quality decision making.

Regulatory agencies focus primarily on protecting patients from exposure to dangerous or ineffective new treatments, historically assuming a position of "protective paternalism." This position presented the second barrier to the use of a more formal approach to B-R assessment. Because B-R decisions are by their very nature subjective, moving from an assessment of discrete information about safety and efficacy to a holistic valuation of the B-R balance requires a decision maker to make a judgment of, or apply value to, the relevance and importance of the benefits and risks. A more formal, systematic approach to B-R assessment would not only improve the transparency of the overall decision-making process, but also offer increased transparency of the decision maker's value judgments. Plainly put, a transparent, systematic process would promote discussion of whether the decision maker's assessments were truly aligned with culturally legitimate and socially representative valuations, thereby undermining the validity of paternal edict.

In early 2007, the FDA withdrew the marketing authority of Zelnorm (tegaserod) for safety reasons. Tegaserod was a drug indicated for the treatment of irritable bowel syndrome, a non-life-threatening but highly debilitating illness. A retrospective review of trial data found a 10-fold increase in the relative risk of significant pooled cardiovascular events, at a rate of 0.1% for tegaserod-treated patients compared with 0.01% for the placebo group. The FDA concluded that because tegaserod was used for a non-life-threatening condition, the 10-fold increase in relative risk of serious cardiovascular events was disproportionate to any potential benefit of the drug. To patients, the FDA's decision appeared to discount the benefit of the potential improvement in their quality of life. The FDA's decision regarding tegaserod stood in contrast with its decision regarding Viagra (sildenafil), a drug widely used to treat erectile dysfunction. Shortly after its introduction, sildenafil was also associated with significant cardiovascular morbidity; however, it remained available while investigations into causality were conducted. Patients and physicians who objected to the tegaserod withdrawal cited the lack of treatment options in a chronic, burdensome disease.[17] The FDA eventually reconsidered its withdrawal decision regarding tegaserod, and responded to patient and physician complaints by creating a restricted access program for patients who had no therapeutic alternatives or who were satisfactorily being treated with tegaserod prior to withdrawal. However, the withdrawal of tegaserod highlighted the FDA's application of a different valuation of benefit and risk than patients and physicians. Why and how had the FDA deemed the cardiovascular risk unacceptable for tegaserod but not sildenafil?

8.6 Regulatory and Industry Benefit-Risk Initiatives: The Rise of Frameworks

With the case for change mounting from multiple sources, the pharmaceutical industry and regulatory agencies began to find themselves transforming their thinking and practice of B-R assessment. With the acknowledgment that past quantitative approaches had failed to be widely adopted, and with the recognition that qualitative processes must precede quantification, they began to consider more structured conceptual approaches to B-R assessment. To improve regulatory decision making and the communication of the overall B-R profile to stakeholders, industry representatives began developing the concept of a B-R framework to form the basis for a common B-R assessment process.

The use of a stepwise process, or framework, to support decision making within federal agencies was not a foreign concept. Other agencies, including the Environmental Protection Agency (EPA), the National Highway Transportation Administration, and the Fish and Wildlife Service, already made use of decision-making frameworks. Each of these agencies, with varying degrees of rigor, utilized qualitative and quantitative methods to evaluate complex concepts, including mortality and morbidity consequent to action or inaction. The EPA, partly in response to the Clean Air Act Amendment of 1991, has been a recognized leader in the field of creating and applying standards for assessing human health impact.

In the United States, a collaborative group of industry scientists working under the umbrella of PhRMA began a 5-year project to develop a structured approach to B-R assessment. The group came to be known as the Benefit-Risk Action Team (BRAT). As had been recommended in the CIOMS report and the EMA reflection paper, the BRAT's goal was to improve the transparency, reproducibility, and ease of communication of B-R assessments of medicines by means of a common framework.

The result of the BRAT's development effort was the PhRMA BRAT Framework for Benefit-Risk Assessment or, as it came to be more commonly known, the BRAT Framework. In this context, a framework for B-R assessment can be thought of as a formal, stepwise process that serves as a common guide to informed B-R decision making. It is a procedure for decomposing and reconstructing a problem that enables decision makers to understand and appreciate the primary drivers of benefit and risk. The framework approach recognizes that a structured and systematic process plays an essential role in assisting and improving human decision making by promoting the clear communication of relevant issues in a transparent, rational, and consistent manner. Although created by representatives within the pharmaceutical industry, the BRAT Framework was developed to serve as a flexible approach that could be applied from the perspective of any stakeholder group—including regulators, payers, healthcare providers, and patients—across the life cycle of a drug.

The BRAT Framework is composed of six steps focused on defining the decision context and selecting, organizing, evaluating, and displaying relevant B-R information for a product, for the purpose of facilitating a summary B-R assessment.[1] A description of the BRAT Framework is given in Table 8.1.

The applicability and generalizability of the BRAT Framework was tested over the course of its development using several scenarios with hypothetical drugs. The framework was found to be effective in assisting, and more importantly improving, B-R decision making and communication. The framework has also been used independently by BRAT members in various applications and public meetings, including FDA advisory committee meetings. As a final step in its development, PhRMA sponsored a BRAT Framework pilot program among its member companies to evaluate its real-world utility and flexibility.[18]

The results of the pilot program were impressive. A full 100% of the pilot participants considered the framework to be useful and flexible, citing applicability to a host of drug development decisions ranging from pre- to post-marketing activities. The framework's primary visualization tools (i.e., key benefit-risk summary (KBRS) tables, value trees, and forest plots) were considered particularly valuable. Participants conveyed that these elements of the framework demonstrated the value of a structured approach to B-R assessment in providing a clear, reproducible organization and summary of complex information in a way that promoted discussion and shared understanding.[19]

Beyond serial testing and the PhRMA pilot, the BRAT was also evaluated as part of the Pharmacoepidemiological Research on Outcomes of Therapeutics by a European Consortium (PROTECT) research program.[20] At the time of this writing, many of the PROTECT publications describing the B-R research activities and results are either still in development or in press. However, preliminary reports have indicated that frameworks are important tools to help govern the B-R assessment process and ensure transparency. The PROTECT work stream devoted to B-R assessment and evaluation found

TABLE 8.1

Description of the BRAT Framework

Step	Description
1. Define the decision context	• Define the drug, dose, formulation, indication, patient population, comparator(s), time horizon for outcomes, and perspective of the decision makers (e.g., regulator, sponsor, patient, physician, or payer).
2. Identify and select benefit and risk outcomes	• Identify and select all relevant outcomes and create an initial value tree.[*] Derived from decision analysis, a value tree (or equivalently, attribute tree) is an explicit visual map of the attributes, in this case the outcomes, of importance to decision makers. • Define a preliminary set of measures for each outcome included in the value tree. • Document the rationale for outcomes included in and excluded from the initial value tree.
3. Identify and extract source data	• Assess data sources relevant to the decision. Determine which of these data will be included in the benefit-risk assessment and briefly document the rationale behind the data source selection. • The data source table is a repository of all the data used to build the framework. Populate the data source table with relevant data, including detailed references and annotations to support subsequent interpretations as appropriate. • As needed or desired, calculate summary measures from information in the data source table for input into a data summary table.
4. Customize the framework	• Modify the value tree based on further review of the data and clinical expertise. • Refine the outcomes and measures. This step may include removing outcomes considered irrelevant to a particular benefit-risk assessment (e.g., to a particular stakeholder group).
5. Assess outcome importance	• If appropriate, apply or assess ranking or weighting to individual outcomes according to the perspective of the decision makers (or other stakeholders).
6. Display and interpret key benefit-risk measures	• The most important summarized representation of data in the BRAT Framework is the key benefit-risk summary (KBRS) table. The KBRS is a flexible table that summarizes the key information needed to quantify outcomes in the value tree. • Present data via visualizations (e.g., graphical displays, forest plots, etc.). • Review summary measures and source data. Identify and address information gaps. • Interpret summary information regarding comparison of benefits and risks; determine whether further data extraction or displays are needed. • Conduct sensitivity analyses to assess the impact of uncertainty in data sources.

Source: Levitan, B., et al., *Clinical Pharmacology and Therapeutics* 89: 217–224, 2011.

[*] The BRAT used the term *value tree* in the sense of values as intrinsic characteristics of benefit and risk. Another term in the decision analytic sense would be *attribute tree* or *attribute hierarchy*, as the outcomes presented in the tree are attributes of a treatment.

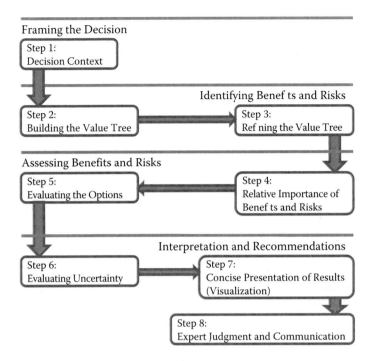

FIGURE 8.1
The CIRS unified benefit-risk framework.

the BRAT Framework well-suited to the tasks and needs of pharmaceutical B-R assessment.[21]

Upon conclusion of the pilot program, the BRAT Framework was transferred to CIRS for further development. In collaboration with B-R assessment experts and regulators, CIRS evaluated the BRAT Framework in light of the framework it employed in its COBRA initiative.[22] This evaluation resulted in a revised framework that emphasized the assessment of uncertainty and included options for weighting benefits and risks (see Figure 8.1).[23] Although weighting remains contentious due to perceived complexity, lack of agreement on appropriate methods, concern over inherent subjectivity, and the specter of undesired or unintended consequences of increased transparency, it was nonetheless included as a formal way to assess the relative importance of individual benefits and risks.

While the PhRMA BRAT initiative was underway in the United States, regulators in other parts of the world were also addressing the framework question. Canadian regulators found that their constituents wanted a modernized drug regulatory system, with a patient-centric approach to regulatory decision making based in science, but socially relevant and responsible. Canadian patients wanted a system that could both protect and promote health through measures that:

- Optimize benefits and minimize harms (in preference to the term *risks*) in the use of drugs
- Optimize patient decision making about whether, when, and how to use drugs
- Allow for alignment of regulatory activities with other partners (decision makers) in the healthcare system in Canada (e.g., manufacturers, payers, healthcare practitioners, and patients)

To support and optimize patient decision making, Canadian regulators and patient groups developed the concept of a patient treatment decision guide. Envisioned as a stepwise, guided process to help patients understand their preferences about possible positive effects, negative effects, and uncertainties of treatment options, the guide is heavily informed by the PrOACT-URL framework.[24] The PrOACT-URL framework (short for Problem, Objectives, Alternatives, Consequences, Trade-Offs, Uncertainties, Risk Tolerance, and Linked Decisions) has been used in other fields, particularly operations research.[25] It has only recently been applied to pharmaceutical B-R assessment.

With respect to Europe, one of the stated goals of the EMA is to make its opinions on the B-R profile of a medicine as consistent and transparent as possible. Like the BRAT, CIRS, and Health Canada initiatives, the EMA recognized there was no standard methodology available to aid the agency in making its B-R decisions. The EMA reflection paper[9] on B-R assessment was followed up by the Benefit-Risk Methodology Project.[26] The Benefit-Risk Methodology Project was started in 2009 to develop and test tools and processes for use in regulatory decision making. Project results included a recommendation to use the PrOACT-URL framework, supplemented as necessary by other methods.[27] The steps of PrOACT-URL are described in Table 8.2.

Step 4 of the PrOACT-URL framework suggests creating an effects table. Conceptually similar to the KBRS table of the BRAT Framework, an effects table displays all the favorable and unfavorable effects that reviewers consider important to the B-R balance, along with definitions of the effects, the unit of measurement for each effect, the plausible range of data, and the measured data (pooled or separate for each clinical trial) associated with the medicinal product and any comparators, including confidence intervals where appropriate. A final column in the effects table is reserved for brief comments about uncertainties.[27]

PrOACT-URL, like the BRAT Framework, was evaluated positively by PROTECT. Under a robust evaluation process inclusive of multiple case studies, both PrOACT-URL and the BRAT Framework emerged as valuable methodologies, with all of the case study participants indicating that the framework structures provided useful guides for the B-R assessment process.[28]

In 2009, with the EMA Benefit-Risk Methodology Project, BRAT initiative, PROTECT, CIRS collaboration, and other country-specific initiatives

TABLE 8.2

Description of the PrOACT-URL Framework

Step	Description
1. Problem	• Determine the nature of the problem, its context, and frame the problem.
2. Objectives	• Establish objectives that indicate the overall purposes to be achieved. • Identify criteria for favorable (benefit) and unfavorable (risk) effects.
3. Alternatives	• Identify the options to be evaluated against the criteria.
4. Consequences	• Describe how the alternatives perform for each of the criteria (e.g., the magnitudes of all effects, their desirability or severity, their incidence). • Create an effects table.
5. Trade-offs	• Assess the balance between favorable and unfavorable effects.

For steps 1–5, only issues concerning the favorable and unfavorable effects, and their balance, have been considered. The next three steps assist in the consideration of how the benefit-risk balance is affected by accounting for uncertainties.

Step	Description
6. Uncertainty	• Assess the uncertainty associated with the favorable and unfavorable effects. • Consider how uncertainty affects the balance by conducting sensitivity analyses and scenario analyses on the model.
7. Risk tolerance	• Judge the relative importance of the decision maker's risk attitude for this product and indicate how this affects the balance reported in step 5 (trade-offs).
8. Linked decisions	• Consider the consistency of this decision with similar past decisions, and assess whether making this decision could have an impact on future decisions.

underway, the FDA began working on a framework of its own. According to the FDA, the Center for Drug Evaluation Research (CDER) identified the need for a more structured B-R assessment in the review process that would help the agency maintain a better line of sight on the overall B-R assessment of drugs during the review process and better communicate the reasoning behind regulatory decisions.

In developing its framework, the FDA examined formal quantitative methods but ultimately concluded that the use of quantitative methods might both force the reduction of complex considerations into a single summary statistic and obscure subjective expert judgment. This led FDA to conclude that a structured qualitative approach, flexible enough to accommodate more complex supporting quantitative analyses when needed, designed to aid rather than replace expert judgment, would best suit its needs. The FDA framework is presented in Table 8.3.

The first two decision considerations or factors listed in Table 8.3, analysis of condition and current treatment options, represent considerations for the broader therapeutic area. Similar to elements embedded in the BRAT

TABLE 8.3

FDA Benefit-Risk Framework

Decision Factor	Evidence and Uncertainties	Conclusions and Reasons
Analysis of condition		
Current treatment options		
Benefit		
Risk		
Risk management		
Benefit-Risk Summary Assessment		

Framework's decision context and the problem step of the PrOACT-URL framework, the analysis of condition and the summary of treatment options are intended to both articulate the medical need for the drug under consideration and summarize current knowledge regarding the condition and relevant therapies. For the analysis of condition, the reviewer is asked to consider the clinical manifestations of the condition, what is known about its natural history, and how the severity of the disease might vary across various subpopulations. For medical need, the reviewer is asked to consider other therapies used to treat the condition, including approved and off-label pharmacological therapies and nonpharmacological therapies. Reviewers are also asked to consider how effective and how well tolerated these alternatives are, and what evidence is available to support the assessment.

Information related to the benefit, risk, and risk management decision factors is specific to the drug under review. For benefit, the reviewer is asked to describe the trials conducted to establish efficacy, including their strengths and weaknesses. Other considerations include endpoints evaluated, the clinical relevance of those endpoints, and whether or not the benefits varied across subpopulations. For risk, the reviewer is asked to characterize safety concerns identified in the clinical trials, including incidence of specific risks in the study population and whether or not the incidence varied by subpopulation. Other questions to consider include: Was there a range in the severity of the risks? Did the risks change with continued exposure? Were the risks reversible when treatment was stopped? Further considerations include how the incidence of any adverse events might change once a drug is available to a wider group of patients in a post-market setting.

Taken together, the first five decision factors in Table 8.3 influence considerations for labeling and risk management activities, as well as potential post-marketing requirements and commitments. The final row of the framework is reserved for the B-R summary assessment. Here, the agency expects review teams to pen a succinct, well-reasoned summary of the FDA's rationale for its regulatory action, including any important clinical judgments that contributed to its decision. Differences of opinion within a review team with respect to scientific or clinical judgment are to be noted in the assessment

> 1. Define the d cision context
> 2. Identity outcomes for consideration
> 3. Data evaluation and summarization
> 4. Interpretation and communication of the assessment

FIGURE 8.2
Core elements of benefit-risk assessment frameworks.

along with an explanation of how those differences were resolved or taken into account in the final decision.[29,30]

8.7 Whose Framework?

It is evident that the use of a framework should be the preferred starting point for a B-R assessment. As explicitly stated in Chapter 5, the application of any quantitative method must be based on a qualitative framework. The question, particularly for industry, is *which* framework? Given the apparent disconnect between the FDA's and the EMA's frameworks, and the considerable role regulatory authorities play in the drug development process, it is not surprising that some companies would wait for specific guidance from regulators before deciding how to proceed. Such delays, however, would be ill-considered as careful review and distillation of each prominent framework reveals common, core concepts required for any useful B-R framework. These are shown in Figure 8.2.

In addition to these core concepts, an analysis of the BRAT, PrOACT-URL, CIRS, and FDA frameworks reveals that they share the following common features:

- Adherence to quality decision-making precepts:
  - Logical soundness
  - Consistency
  - Practicality
  - Transparency
- Use of systematic approaches with embedded decision aids to structure evaluations:
  - Identification of which benefits and risks are considered most relevant
  - Evaluation of the quality and interpretation of evidence
  - Transparency of how the benefits and risks are weighted or prioritized

- A focus on questioning what can be done to optimize the B-R balance by managing and mitigating the risks
- Use of a decision context with recognition that useful B-R assessments depend on:
  - Indication of the treatment and severity of the condition
  - Medical need
  - Population(s) being treated
  - Availability of other available treatments
  - Perspective from which the decision is being made (e.g., regulator, sponsor, patient, payer, clinician)
- Use of a B-R profile that consists of defining benefit and risk criteria, noting the measures for the criteria, and inclusive of documentation of the rationale for each criterion's inclusion or exclusion in the assessment
- Evaluating, summarizing, and communicating relevant data and information via tools such as:
  - Value trees
  - Effects tables
  - Key benefit-risk summary table
- Tracking uncertainties
- Communication of the rationale underpinning the assessment and its potential impact on subsequent decisions

Recall the requirements for a B-R assessment process articulated by the CIOMS and IOM reports, in addition to the EMA reflection paper: the core elements of the frameworks discussed meet all of the stated requirements. The BRAT Framework, CIRS framework, and PrOACT-URL will all reward their users with a robust process and relevant outputs that can be customized and packaged according to the needs and requirements of their customers, whoever they may be.

References

1. Coplan, P., Noel, R., Levitan, B., Ferguson, J., and Mussen, F. 2011. Development of a framework for enhancing the transparency, reproducibility and communication of the benefit-risk balance of medicines. *Clinical Pharmacology and Therapeutics* 89: 312–315.
2. Levitan, B., and Mussen, F. 2012. Evaluating benefit-risk during and beyond drug development: An industry view. *Regulatory Rapporteur* 9: 10–15.

3. Mussen, F., Salek, S., and Walker, S. 2009. *Benefit-risk appraisal of medicines: A systematic approach to decision-making.* Chichester, UK: John Wiley & Sons.

4. CIOMS Working Group IV. 1998. *Benefit-Risk balance for marketed drugs: Evaluating safety signals.* Geneva: Council for International Organizations of Medical Sciences.

5. Walker, S., McAuslane, N., and Liberti, L. 2011. *Developing a common benefit-risk assessment methodology for medicines—A progress report.* Scrip Regulatory Affairs 18–21.

6. Kaufman, M. 2006. FDA is criticized over drugs' safety problems. *Washington Post,* April 24, 2006.

7. Committee on the Assessment of the US Drug Safety System, Baciu, A., Stratton, K., and Burke, S., ed. 2007. *The future of drug safety: Promoting and protecting the health of the public.* Washington, DC: National Academies Press.

8. *Drug safety: Improvement needed in FDA's postmarket decision-making and oversight process.* Report GAO-06-402. March 2006. Washington, DC: Government Accountability Office.

9. European Medicines Agency. 2007. *Report of the CHMP working group on benefit-risk assessment models and methods.* http://www.ema.europa.eu/docs/en_GB/document_library/Regulatory_and_procedural_guideline/2010/01/WC500069668.pdf.

10. Centre for Medicines Research. 2005. *Benefit-risk assessment model for medicines: Developing a structured approach to decision making.* Report of the workshop organized by the CMR International Institute for Regulatory Science, Washington, DC, June 13–14, 2005.

11. Felli, J., Noel, R., and Cavazzoni, P. 2009. A multiattribute model for evaluating the benefit-risk profiles of treatment alternatives. *Medical Decision Making* 29: 104–115.

12. Centre for Innovation in Regulatory Science. 2011. *Visualizing benefit-risk of assessment of medicines: The key to developing a framework that informs stakeholder perspective and clarity of decision making. Workshop report.* Washington, DC, June 16–17, 2011. http://cirsci.org/system/files/private/1037_CIRS_June_WS_synopsis_0.pdf.

13. Directive 2001/83/EC of the European Parliament and of the Council of 6 November 2001 on the community code relating to medicinal products for human use. http://www.edctp.org/fileadmin/documents/ethics/DIRECTIVE_200183EC_OF_THE_EUROPEAN_PARLIAMENT.pdf.

14. Tallarida, R., Murray, R., and Eiben, C. 1979. A scale for assessing the severity of diseases and adverse drug reactions: Application to drug benefit and risk. *Clinical Pharmacology and Therapeutics* 25: 381–390.

15. Gelber, R., Cole, B., Gelber, S., and Goldhirsch, A. 1995. Comparing treatments using quality-adjusted survival: The Q-TWiST method. *American Statistician* 49: 161–169.

16. Chuang-Stein, C. 1994. A new proposal for benefit-less-risk analysis in clinical trials. *Controlled Clinical Trials* 15: 30–43.

17. Brandt, L. 2008. The FDA's decision-making process: Isn't it time to temper the principle of protective paternalism? *American Journal of Gastroenterology* 103: 1226–1227.

18. Levitan, B., Andrews, E., Gilsenan, A., Ferguson, J., Noel, R., Coplan, P., and Mussen, F. 2011. Application of the BRAT Framework to case studies: Observations and insights. *Clinical Pharmacology and Therapeutics* 89: 217–224.
19. Noel, R., Hermann, R., Levitan, B., Watson, D., and Van Goor, K. 2012. Application of the Benefit-Risk Action Team (BRAT) Framework in pharmaceutical R&D: Results from a pilot program. *Drug Information Journal* 46: 736–743.
20. Innovative Medicines Initiative. PROTECT. http://www.imi-protect.eu/
21. Nixon, R. 2012. Benefit-risk analysis: Experience in using the BRAT Framework. Presented at PSI Conference, Birmingham, UK, May 15, 2012.
22. Centre for Innovation in Regulatory Science. UMBRA update. June 15, 2012. http://213.120.141.158/sites/default/files/UMBRA%20Issue1,%202012.pdf.
23. Centre for Innovation in Regulatory Science. UMBRA update. February 2013. http://213.120.141.158/sites/default/files/CIRS%20UMBRA%20Issue%20 2%20Feb%202013.pdf
24. Lim, R. 2013. Modernizing the drug regulatory landscape in Canada to benefit patients: An update on benefit-harm-uncertainty. Presented at Drug Information Association webinar series Assessing the Benefits and Risks of Medicines: A Changing Regulatory Landscape, January 30, 2013.
25. Hammond, J., Keeney, R., and Raiffa, H. 1999. *Smart choices: A practical guide to making better decisions.* Boston: Harvard University Press.
26. European Medicines Agency. *Benefit-risk methodology project.* http://www.ema. europa.eu/ema/index.jsp?curl=pages/special_topics/document_listing/document_listing_000314.jsp&mid=WC0b01ac0580223ed6.
27. European Medicines Agency. *Benefit-risk methodology project work package 4 report: Benefit-risk tools and processes.* http://www.ema.europa.eu/docs/en_GB/document_library/Report/2012/03/WC500123819.pdf.
28. Ashby, D. 2012. What role should formal risk-benefit decision-making play in the regulation of medicines? Presented at 32nd Conference on Applied Statistics, Dundalk, Ireland, May 16–18, 2012.
29. Frey, P. 2012. CDER's benefit-risk framework. Presented at the Drug Information Association North American Annual Meeting, Philadelphia.
30. Food and Drug Administration. 2013. Structured approach to benefit-risk assessment in drug regulatory decision-making. Draft PDUFA V implementation plan. February 2013. http://www.fda.gov/downloads/forindustry/userfees/prescriptiondruguserfee/ucm329758.pdf.

Section IV

Post-Launch Assessment

Introduction

The door to the Post-Launch Assessment gallery is new and slides open easily at a gentle touch. The chamber beyond is under construction. So too are the works of art that lay scattered about on tables and easels—sketches in graphite and charcoal, chalk pastels, paintings in watercolors and oils. All are unfinished, some scarcely started.

There are photographs clipped to some of the easels, and a few more scattered about the tables. They appear to serve as models for the artists. Many of the photographs bear familiar images—works on display in the Full Clinical Development gallery—though a few depict unfamiliar objects. The artists' renditions, though unfinished, range from near-perfect reproductions to deliberate distortions of their subjects. Some seek to reproduce what they have observed; some experiment with modifications. Some work without a photograph. Some work from imagination alone.

9

Considerations and Strategies for Benefit-Risk Assessment in the Real-World Setting

Alicia Gilsenan and Elizabeth Andrews

CONTENTS

9.1 Introduction

9.1.1 Why Bother?

If you are in a pharmaceutical or biologic company and have done a prelaunch benefit-risk (B-R) assessment, why would you bother updating the B-R assessment post-launch? Now that the drug is on the market, what is the point?

The most straightforward answer is that it is a regulatory requirement now that the European Medicines Agency (EMA) has released Module VII of the EMA *Guidelines on Good Pharmacovigilance Practices* (GVP),[1] which explicitly requires inclusion of an integrated B-R assessment in periodic safety update reports (PSURs). Prior to this guidance, companies were required to submit a formal assessment of each identified and potential risk as part of their risk management plan for each product, as well as a PSUR. Both of these documents focused primarily on risks. However, even without this regulatory requirement, it has already become a practice for selected products for some companies to examine a product's B-R assessment post-launch. Such assessments can provide a strategic product advantage, particularly for any product that may face public comparisons with other products when the market landscape changes following initial product approval, or when important new safety signals emerge unexpectedly.

Conducting a new B-R assessment, or updating a preapproval assessment using information available post-launch, may seem unwise for a number of reasons. For example, a new assessment may require significant internal company resources, may present large opportunity costs consequent to deploying valuable resources to the analysis, and may carry with it the possibility of uncovering important information gaps or unfavorable comparative information. Some companies may consider the additional work triggered by initial findings from these assessments to be unproductive at best.

In addition to these concerns, for many companies a post-launch B-R assessment is an unfamiliar and uncomfortable process for members of clinical teams whose "comfort zones" lie in the realm of randomized controlled clinical trials. The post-launch B-R assessment process may require exploration of information sources that may be poorly constructed or inadequately described, such as observational studies of benefits and risks. Summarizing data on outcomes across studies and converting disparate measures to common metrics will likely be challenging due to statistical complexity and other uncertainties. What is the best approach to overcome these challenges? With appropriate internal and external expertise, and a healthy appreciation of the nature of observational data, post-launch B-R assessments are likely to be seen as valuable, perhaps even indispensable.

Many products have been introduced to the market based on excellent clinical trials data, only to be challenged later because of information that emerged post-launch. For example, newly emerging information on the risk of serious adverse events forced two drugs, natalizumab for multiple sclerosis[2] and alosetron for irritable bowel syndrome,[3] to be removed from the market, subjected to reevaluation of benefits and risks, and relaunched with restrictive risk management programs. By contrast, bupropion, initially approved for depression in 1986, was withheld from the market following approval because of an increased risk of seizures observed in clinical trials. Following the conclusion of a large safety study, the drug was reapproved and commercially launched in 1989. Despite an acknowledged increase in risk of seizure, the benefits of the treatment were viewed to exceed this risk, for clinical depression as well as for a newer indication, smoking cessation.

The process of developing the B-R assessment (particularly itemizing key risks and benefits, and visually displaying differences between the target product and relevant comparators) is inherently constructive and helps prepare a sponsor company for future regulatory discussions about its product (e.g., new indications, safety signals, changes in the competitive landscape). While the B-R assessment process may uncover evidence needs that must be pursued, it will also enable existing data/evidence to be better organized, understood, and communicated. This, in turn, will facilitate productive discussions within the company and with external stakeholders, especially regulators. The organization and transparency of the evidence expedites creation of displays for alternative scenarios and enables sensitivity analyses to explore uncertainty about the data—for example, detection of a serious adverse event based on spontaneous reports with no firm estimate of the occurrence rate. Aggregated data displays, particularly those demonstrating benefit and risk differences, are very helpful in putting signals of rare serious events into context because they show the population impact (e.g., large relative risk, but extremely low background rate).

Making B-R judgments is not only important for pharmaceutical or biologic companies, but also an ongoing function of drug regulatory agencies. Although this chapter is focused on assessments performed by sponsoring companies, the same methodological principles apply to assessments performed by regulatory agencies and other stakeholders.

If we are now convinced of the inherent value in conducting B-R assessments post-launch, or at least the necessity of doing so, we may ask what else there is to consider that is unique to the post-approval setting and how to conduct a useful and informative B-R assessment.

9.1.2 What Is Different from Prelaunch Benefit-Risk Assessment?

B-R assessments conducted before launch have the luxury of resting largely on data from well-controlled randomized clinical trials. Most licensure trials are designed to optimize one's ability to measure efficacy, if it exists, while minimizing the number of individuals subjected to experimental conditions. Randomization reduces the impact of systematic bias, selection of relatively homogeneous and uncomplicated study subjects reduces influence of extraneous factors (e.g., comorbid conditions, other treatments), and close monitoring helps ensure high compliance with the treatment regimen and collection of high-quality data. Such trials may demonstrate efficacy in the limited study population and provide proof of concept; however, they rarely demonstrate population effectiveness. In contrast, post-market B-R assessments must account for real-world circumstances of medication use and incorporate information from observational studies, as well as account for potential risks that may have been identified through spontaneous reports. Key differences are highlighted below.

9.1.2.1 Population

The population actually using a medication must be taken into consideration and will inevitably include special subpopulations not included in clinical trials (i.e., children, pregnant women, elderly, those taking multiple drugs, or those with multiple comorbidities). Off-label use or misuse can have an impact on a B-R assessment, as demonstrated by the "fen-phen" example— the rapid market withdrawal of the diet drugs fenfluramine and dexfenfluramine due to an increase in heart valve defects[4] was thought to be heavily influenced by widespread prescribing of the combination of fenfluramine and phentermine (known as fen-phen) to patients, including some who were only mildly overweight. According to EMA guidance, estimates and potential consequences of off-label use and misuse must be addressed in risk management plans and PSURs. Interestingly, risks from unauthorized use must be included in the PSUR B-R assessment, but benefits can be included only for authorized (i.e., label) indications.

9.1.2.2 Compliance/Adherence

Data from clinical trials are usually analyzed as "intent to treat," an approach that analyzes all patients according to the treatment groups to which they were assigned, regardless of compliance/adherence. This approach is required to preserve the benefits of randomization. However, at a population level, we are interested in the effectiveness and safety of products based on actual use. In observational studies, compliance/adherence is rarely measured with high accuracy but can have a significant impact on the effectiveness and side effect profile of a medication. An intent-to-treat analysis is rarely appropriate in such studies. Therefore, it is important to consider how compliance/adherence is measured and analyzed in the studies included in a B-R assessment. Observational studies that do not take compliance or adherence into account may over- or underestimate the true effects of a medication. For example, a meta-analysis of medications used to prevent cardiovascular events showed that only 57% of patients maintained adherence after a median of 24 months, and adherence continued to decline over time.[5] If a benefit demonstrated in clinical trials were applied to a real-world population where only 57% of the patients proved adherent, the benefit would be overstated.

9.1.2.3 Comparator

Comparators selected for clinical trials are usually limited in number and type to optimize the trial design consistent with scientific and ethical principles. These comparators may be limited to placebo, usual care, or a single alternative treatment. Post-launch, comparisons relevant to physicians and patients will include all available alternatives, including those newly entering the market, in a context of changing patterns of care and evolving prescribing patterns. Indeed, a B-R assessment may yield different results over time with the emergence of new alternatives. For example, consider isotretinoin, a highly effective treatment for acne that causes birth defects in a high percentage of exposed pregnancies. How likely would it be for isotretinoin to remain available if an equally effective medication were introduced to the market that did not carry such a risk?

9.1.2.4 Stakeholders

Stakeholders for B-R assessments in a real-world setting are more diverse than at time of approval, when regulators can focus only on the submission package and cannot consider factors such as cost. As the audience for post-launch B-R assessment information broadens to include prescribers, other government entities and payers, patients, and sometimes the media, the information needs widen. In addition, regulators may shift their viewpoints over time. Two examples mentioned earlier (natalizumab for multiple sclerosis and alosetron for irritable bowel syndrome) were withdrawn from the market due to potential severe side effects. In both cases, patients contacted the U.S. Food and Drug Administration (FDA) as well as the sponsor pharmaceutical companies to request access to the withdrawn medications due to their personal perspectives that the benefits they received outweighed the risks they faced. After a review of safety data and the development of restrictive risk minimization programs, both drugs were returned to the market. Had patients never been exposed to the medications in a post-approval setting, and had they not advocated for the products' availability, it is highly unlikely that the drugs would have ever been reapproved.

9.1.2.5 Indications

Treatment indications, formulations, dosages, and patterns of prescribing and use may vary by country, depending on the approved product and product information, as well as many other factors, such as inclusion in health plan formularies, reimbursement status (by the National Institute for Health and Care Excellence (NICE) in the United Kingdom, for example), and restrictions on use. Such diversity in availability, access, and use must be considered in evaluating results from observational studies conducted in different geographic areas. In addition, new or expanded indications, which may have an impact on a product's overall B-R assessment, may be approved after its initial launch.

9.2 Special Design Considerations in the Post-Launch Setting

9.2.1 What to Measure

Outcomes measured in a post-launch B-R assessment need to include key outcomes measured in registration clinical trials, but must go beyond those in several ways. Benefit outcomes of interest derived from clinical trials may need to be redefined in a way that reflects actual use, compliance, and real-world patient characteristics. Benefit outcomes of interest may also need to be expanded to include those not routinely measured in clinical trials but considered important by patients and prescribers, such as quality of life, tolerability, and patient preferences. Other benefit outcomes may emerge from anecdotal observation that may require confirmation in formal studies. For example, the relative side effect profiles of antidepressants (e.g., effects on weight and libido) are important to patients and prescribers when choosing a treatment. Safety outcomes selected for a post-launch assessment will need to include outcomes that emerge from randomized and observational safety studies, spontaneous adverse event reports, and published case reports. Unexpected safety events may be reported, and events previously observed in clinical trials may appear with a different frequency or severity when the product is in actual use. Some safety outcomes considered to be possible signals in clinical trials based on laboratory results may not be confirmed in subsequent studies. For example, a small increase in blood pressure seen in clinical trials may not translate into an increase in serious cardiovascular events observed in large observational studies. Serious but rare outcomes pose a challenge to quantitative and semiquantitative methods: convincing case reports coupled with biological plausibility may be sufficient to conclude a causal association without any estimate of the frequency of occurrence. In the absence of reliable estimates, the incidence of an outcome may be projected at different levels to evaluate the sensitivity of B-R conclusions to these levels.

9.2.2 Sources of Information

Information from preapproval trials needs to be supplemented with information from other sources in the post-launch phase, including new or extended clinical trials, observational studies, published systematic literature reviews and meta-analyses, and case reports (including published cases and case series, as well as spontaneous adverse event reports). Combining and synthesizing information across such disparate sources requires careful evaluation of all underlying sources, thoughtful selection of the appropriate measures, and meticulous treatment of statistical uncertainty. These issues are commonly encountered in conducting meta-analyses and comparative effectiveness review.

TABLE 9.1

Key Considerations for Benefit and Risk Data Sources

Attribute	Benefits	Nonserious Risks	Serious Risks
Typical data sources	Phase 3 trials Post-marketing trials	Phase 3 trials Post-marketing trials	Observational studies Safety trials Spontaneous reports (not applicable)
Indications	One per trial	One per trial	Mixed, if misclassified
Comparator groups	Placebo Standard of care Other treatments	Placebo Standard of care Other treatments	Standard of care No comparator in some studies
Typical time course	<6 months	<6 months	Up to 2 years
Frequency	10–80%	1–20%	<1%
Confidence interval	Narrow	Narrow	Wide
Data quality	Good	Good	Variable
Effect measure	Proportion difference	Proportion difference	Relative risk (risk ratio) Risk difference
Confounding	Minimal	Minimal	Extensive

Much guidance is available regarding meta-analyses and comparative effectiveness reviews;[6–11] however, it remains imperative that researchers conducting B-R assessments appreciate the differences in types of data, sources of data, and their potential to confound when examining evidence related to benefits and nonserious, frequent adverse events compared with rarer, serious events. It may be necessary to transform data from clinical trials to match those collected in observational studies or vice versa to allow for pooling. A few highlights for consideration when synthesizing data for a B-R assessment are provided in Table 9.1.

9.2.3 Selection of Appropriate Comparative Measures

Because of the nature and size of clinical trials, the differences between the experimental and comparator products are usually expressed in relative risk measures. However, when projecting the impact of benefits and risks to populations, it is more relevant to consider absolute risks (e.g., 1 event per 1,000 person-years of exposure) and risk differences (e.g., one additional case or event per 1,000 person-years of exposure). Otherwise, the influence of the baseline risk of events can be masked by relative risk measures: for instance, a relative risk of 2 has vastly different impact on a population for a condition that occurs frequently (1 in 10) than one that occurs rarely (1 in 100,000). Another issue that arises in comparing results across multiple studies is that outcomes that were measured in studies as continuous measures (e.g., lipid levels) may need to be converted to a different but corresponding measure (e.g., the percentage of patients meeting a threshold target).

9.2.4 Confounding

Confounding by indication, or channeling bias, is the most challenging methodological issue encountered in evaluating the results of observational post-approval studies of safety and effectiveness. In contrast to treatment randomization in a trial, the factors that influence physicians and patients to select certain medications may be independently associated with an effectiveness or safety outcome of interest. For example, a new treatment may be introduced initially to patients who have the poorest clinical prognosis and are already at high risk of adverse events. Such selective prescribing must be addressed in observational studies.

Pharmacoepidemiology researchers continue to develop and apply more sophisticated tools to control for such confounding and to estimate the possible effect of unmeasured confounding.[12] These include use of the modeling of factors associated with treatment choice to create propensity scores, use of instrumental variables, and sensitivity and bias analyses. Given the nature of the underlying data used in some observational studies and the inability to directly measure all variables of interest, it is important to be transparent about the level of uncertainty that exists in the information that supports a B-R assessment.

9.3 Benefit-Risk Assessment Methods

All of the methods explored in previous chapters are applicable to the post-launch assessment. A framework for collecting and organizing B-R information can be readily extended to include new sources and types of information. The complexity of assessments will increase with the need to evaluate products against multiple comparators, the availability of new information sources, the complexity of data integration, and emergence of new information such as patient preference data not available prior to approval.

9.3.1 Qualitative

The FDA framework[13] is one example of a strictly qualitative approach. The goal for the development of this framework was to make regulatory decisions more transparent. The framework allows documentation of benefits and risks considered, shows how the evidence was interpreted, and provides a "big picture" look of the overall assessment. See Chapter 8 for additional detail.

9.3.2 Semiquantitative

Two qualitative approaches that may be extended to incorporate quantitative analysis have been developed: the BRAT Framework developed by the Pharmaceutical Research and Manufacturers of America (PhRMA)[14,15] and PrOACT-URL, developed by Hammond et al.[16] and embraced by an EMA working group.[17] See also Chapter 5 for further discussion. Both tools provide guidance for systematically assessing the benefits and risks of a product. For example, the steps for the BRAT Framework begin with first agreeing on the decision frame or context (drug, indication, comparator, population, and perspective for the analysis). Next, benefit and risk attributes are organized in a value tree, forcing clear thought on what known potential benefits and risks are most important and should be included in the assessment. Once the benefit and risk attributes are agreed upon, the measures for each attribute are selected. These steps provide the framework for characterizing and gathering data that will be used to inform a B-R assessment. Using meta-analytic or other techniques, the data can then be pooled and visual displays can be generated. Chapters 1, 5, and 8 provide additional detail.

9.3.3 Quantitative

A number of quantitative methods have been proposed as foundations for B-R analysis, varying widely in scope, complexity, and computational intensity. For the most part, these methods have been developed with an eye toward guiding clinical development and supporting regulatory review. As such, they tend to rely heavily on scientifically relevant measures and hard clinical endpoints. However, given appropriate expansion of their underlying frames and proper treatment of data employed, many are well suited to post-launch B-R assessment. Chapters 5 and 8 provide some exposition and discussion for a number of popular and proposed quantitative methods for B-R assessment.

9.4 Conclusions

Because of the methodological differences between clinical trials and observational data, the resources needed to conduct post-launch B-R assessments must include not only individuals expert in observational methods (i.e., epidemiologists), but also those knowledgeable about the clinical safety outcomes, and individuals familiar with actual use patterns (e.g., market research, medical affairs), in addition to the standard clinical development team. This set of expertise is considerably broader than typical for development teams responsible for preparing traditional periodic safety reports.

The challenges of continuing to evaluate a product's relative benefits and risks post-launch are indeed daunting and could consume far more resources than appropriate, begging the question: What is the B-R balance for conducting post-launch B-R assessments? EMA guidance requires some level of continuing assessment for all products; the extent to which those assessments will be primarily qualitative or quantitative, relatively superficial or extremely robust, remains to be seen following the implementation of the new guidance. However, conducting a formal assessment appears most important at key regulatory milestones, such as the introduction of a new indication, recognition of a major new safety signal, and times of major change in the market landscape.

References

1. European Medicines Agency. 2012. *Guidelines on good pharmacovigilance practices (GVP). Module VII. Periodic safety update report.* EMA/816292/2011. http://www.ema.europa.eu/docs/en_GB/document_library/Scientific_guideline/2012/06/WC500129136.pdf.
2. Food and Drug Administration. 2006. FDA approves resumed marketing of Tysabri under a special distribution program (online news release). http://www.fda.gov/NewsEvents/Newsroom/PressAnnouncements/2006/ucm108662.htm.
3. Food and Drug Administration. 2002. Questions and answers about Lotronex. http://www.fda.gov/Drugs/DrugSafety/PostmarketDrugSafetyInformationforPatientsandProviders/ucm110859.htm.
4. Food and Drug Administration. 1997. FDA announces withdrawal of fenfluramine and dexfenfluramine (fen-phen). http://www.fda.gov/Drugs/DrugSafety/PostmarketDrugSafetyInformationforPatientsandProviders/ucm179871.htm.
5. Naderi, S., Bestwick, J., and Wald, D. 2012. Adherence to drugs that prevent cardiovascular disease: Meta-analysis on 376,162 patients. *American Journal of Medicine* 125: 882–887.
6. Liberati, A., et al. 2009. The PRISMA statement for reporting systematic reviews and meta-analyses of studies that evaluate health care interventions: Explanation and elaboration. *Annals of Internal Medicine* 151: W65–W94.
7. Crowe, B., et al. 2009. Recommendations for safety planning, data collection, evaluation and reporting during drug, biologic and vaccine development: A report of the safety planning, evaluation, and reporting team. *Clinical Trials* 6: 430–440.
8. Ioannidis, J., et al. 2004. Better reporting of harms in randomized trials: An extension of the CONSORT statement. *Annals of Internal Medicine* 141: 781–788.
9. Huang, H., Andrews, E., Jones, J., Skovron, M., and Tilson, H. 2011. Pitfalls in meta-analyses on adverse events reported from clinical trials. *Pharmacoepidemiology and Drug Safety* 20: 1014–1020.

10. Agency for Healthcare Research and Quality. 2012. *Methods guide for effectiveness and comparative effectiveness reviews.* http://www.effectivehealthcare.ahrq.gov/ search-for-guides-reviews-and-reports/?pageaction=displayproduct&mp=1& productID=318.
11. Cochrane Collaboration. 2011. *Cochrane handbook for systematic reviews of interventions.* Version 5.1.0 (updated March 2011). http://www.cochrane.org/training/ cochrane-handbook.
12. Rothman, K., Greenland, S., and Lash, T. 2008. *Modern epidemiology.* 3rd ed. Philadelphia: Lippincott.
13. Frey, P. 2012. *Benefit risk considerations in CDER: Development of a qualitative framework (DIA meeting).* Silver Spring, MD: Food and Drug Administration, Center for Drug Evaluation and Research. http://www.fda.gov/downloads/ AboutFDA/CentersOffices/OfficeofMedicalProductsandTobacco/CDER/ UCM317788.pdf.
14. Coplan, P., Noel, R., Levitan, B., Ferguson, J., and Mussen, F. 2011. Development of a framework for enhancing the transparency, reproducibility and communication of the benefit-risk balance of medicines. *Clinical Pharmacology and Therapeutics* 89: 312–315.
15. Levitan, B., et al. 2011. Application of the BRAT Framework to case studies: Observations and insights. *Clinical Pharmacology and Therapeutics* 89: 217–224.
16. Hammond, J., Keeney, R., and Raiffa, H. 1999. *Smart choices: A practical guide to making better decisions.* Boston: Harvard University Press.
17. European Medicines Agency. 2011. *Benefit-risk methodology project. Work package 3 report: Field tests.* EMA/718294/2011. http://www.emea.europa.eu/docs/ en_GB/document_library/Report/2011/09/WC500112088.pdf.

10

Benefit-Risk Assessment and the Payer Perspective

Joseph Johnston, Ralph Swindle, James Felli, and Don Buesching

CONTENTS

10.1 Introduction

Assessing benefits and risks is at the heart of decisions payers must make regarding access to medications authorized for marketing via the regulatory process. This common foundational use of benefit and risk by payers and regulators in their decision processes suggests an opportunity to harmonize the production and use of evidence regarding benefit and risk, and some evidence of harmonization is already beginning to emerge. However, unlike regulators, payers must also consider the cost of new treatments in their decisional framework, and this difference has sometimes stood in the way of mutual understanding and cooperation between regulators and payers. To better understand the role of benefit and risk in the decisions faced by payers, we will consider their perspective and operating environment, and the role of health technology assessment (HTA) in their decision making. We will then outline recent trends in payers' use of evidence and HTA, and highlight future opportunities for cooperation between payers and regulators.

10.1.1 How Does the Payer Perspective Differ from the Regulatory Perspective?

Regulators have the responsibility to ensure that medications authorized for marketing are both safe and effective for the populations for which they are indicated. Since nearly all medications carry some level of risk in their therapeutic use, regulators need to ensure an appropriate balance between their risks in use and their benefits in use, as discussed in Chapters 7 and 8. Since the early 1960s, randomized controlled trials (RCTs) have emerged as the gold standard for assessing benefits as well as highly to moderately prevalent risks. By the end of the drug development process, regulators generally have a strong, high-quality evidentiary base upon which to base their decision whether or not to grant marketing approval. Payers also rely heavily on RCTs to establish a fundamental understanding of benefit and risk. However, because of the nature of their decisions, they still have several evidentiary gaps that make their decisions less clear at the point of market authorization.

Payers have the responsibility to consider benefit and risk for the population under treatment for a new medication, but also a fiduciary responsibility to those responsible for paying for the medication. In the most expansive situation of a single, national payer system, this turns out to be society at large. This fiduciary responsibility greatly shapes the nature of payer decisions. For a new medication, financial accountability leads payers to ask questions of two broad types: Does the new medication represent good value for money compared with alternative treatments for the same condition? Given the cultural values and preferences for health versus other goods and services, is the new medication affordable? It is quite possible for a medication that provides evidence of value for money to be deemed unaffordable and for access to it to be denied because of its unacceptable anticipated impact on the healthcare budget.

This possibility serves to highlight the nature of medication access and reimbursement decisions made by payers. Payers must translate benefit and risk evidence into considerations of value for money and affordability, and make a series of decisions, including whether access to the medication will be granted at all, whether use is restricted to specific subpopulations or treatment scenarios (e.g., first-line use versus use only after failure of other medications), and what portion of costs to assume directly and what to pass along to patients through copays and other mechanisms. This distinct payer mandate also serves to explain how decisions made by regulators and payers can appear to conflict. A medication judged by regulators to have an acceptable, perhaps even a desirable, benefit-risk (B-R) profile may nonetheless be denied access by payers because of the economic burden it would impose.

10.1.2 The Rise of Real-World Evidence in Payer Decision Making

Because they are responsible for the benefits patients enjoy under usual care treatment, the risks they may encounter there, and the costs that accrue, payers have a clear need for economic information on medications, side effects, hospitalizations, ambulatory visits, and a whole host of other inputs that produce a patient's return to health. In short, payers need information on the real-world practice of medicine because that is where their risk lies in decision making.

So where does this type of information come from? RCTs are an unlikely source. A clinical trial is designed to maximize the likelihood that, other things being equal, the mechanism of action of a new medication is responsible for the outcomes observed in the trial. Controlling study conditions to make other things equal is, in fact, a principal reason why RCTs are considered a gold standard form of experimental design. By contrast, in the real-world practice of medicine, hardly anything can be held equal: patients come from a variety of demographic and cultural backgrounds with a variety of different types of comorbid conditions and concomitant treatments; physicians vary widely in their ability and motivation to treat different types of patients; and health systems can be quite heterogeneous in the way care is delivered and financed. There are a variety of naturalistic, nonexperimental research designs and analyses that can capture useful information after a drug has been introduced into the real-world practice of medicine; however, unlike RCTs, establishing causal inference in such designs can prove quite challenging. Unfortunately for payers, key decisions about access and reimbursement must be made soon after market authorization, before such naturalistic studies can deliver evidence.

The set of tools and processes by which payers use evidence from both RCTs and other sources of real-world information to model how benefit, risk, and cost are likely to play out in real-world treatment settings is coalescing around the common term *health technology assessment* (HTA). In 1985, the U.S. Institute of Medicine defined HTA as

> any process of examining and reporting properties of a medical technology used in health care, such as safety, efficacy, feasibility, and indications for ethical consequences, whether intended or unintended.[1]

10.2 Role of Health Technology Assessment in Payer Decision Making

Payers use HTA to bridge the gap between the essential but limited insights into expected real-world product effectiveness and safety provided by RCT data available at the time of regulatory approval and the more complete understanding that they will eventually derive from an examination of real-world use in their population of interest. While there can be considerable variability in the approaches taken, methodologies employed, and processes by which HTA outputs are applied to access and reimbursement decisions, the basic components of an HTA are generally the same.

10.2.1 What Are the Components of an HTA?

An essential first component of an HTA is an understanding of disease burden and unmet need in the HTA conductor's population of interest. Ideally, this should include an understanding of both disease impact on individual affected patients and the extent of the disease at a population level (i.e., disease epidemiology). While population estimates of disease incidence and prevalence derived from the general medical literature and other sources may serve as a starting point, they are ultimately useful only to the extent that they can influence decision making by informing estimates of expected resource utilization in the payer's population. Specifically, payers are interested in understanding how widely a new drug will be used, whether its use is likely to conform to the labeled indication or expand beyond it, and whether use of the drug will result in an increase or decrease in the use of other expensive technologies and services. All of this contributes to an understanding of expected budget impact and informs the crucial question of affordability.

The formal review of existing data regarding product efficacy and safety comprises a second key component of an HTA. While RCT data provide a solid baseline, payers will also ask whether they can expect the results to translate into real-world effectiveness in their covered populations. Understanding the applicability of a trial's results to a payer's covered population requires an examination of the representativeness of patients enrolled and the treatment received in the trial. Were patients with comorbid medical conditions or those taking other commonly prescribed medications excluded from trials? (Usually yes.) Was the care in these trials provided by highly trained clinical trial investigators working at select specialty clinics? (Usually yes.) How similar was this care to that likely to be provided at general practice sites in the payer's practice environment? (Usually not very.) Additionally, a payer will seek to understand the B-R profile in specific trial subpopulations, and estimate the B-R profile in off-label populations not treated in trials who may nonetheless receive treatment in the real world. This understanding can help to justify the use of access controls to drive use toward patients likely to derive the greatest benefit from treatment.

A third important component is the assessment of the evidence surrounding relevant comparator products, which often extends beyond the comparators used in the trials supporting regulatory approval. Again, because the concern is with the financial impact of the introduction of a new technology into a complex healthcare system, relevant comparators may include not only similar drugs in the same medication class, but also drugs in other classes used to treat the same disease, as well as complementary/alternative treatments (e.g., acupuncture for low back pain) and nonpharmacologic treatments (e.g., psychotherapy for depression). As head-to-head trials comparing a new drug to the full gamut of potentially relevant treatment alternatives are seldom available, HTA assessors must make do with indirect comparisons.

The final critical component of an HTA consists of one or more economic assessments addressing questions of value and affordability. Complete economic analyses (e.g., cost-effectiveness analyses) consider the cost of a technology in the context the benefits delivered, while partial economic analyses (e.g., budget impact analysis) consider cost in isolation.

10.2.2 Tools and Conduct of HTA—How Are Benefit and Risk Extrapolated to the Real World?

The foundation for any HTA is a formal assessment of what is known about the efficacy and safety of a new treatment and relevant comparators based on clinical trial evidence. This begins with a structured examination of data from pivotal clinical trials and a systematic review of the published literature. This information is commonly assembled into evidence tables, akin to the evidence tables and effects tables discussed in Chapters 1 and 5. While the recommended or required format for these tables can be profoundly different from one requestor to the next, the content is basically the same: a description of trial characteristics and results in sufficient detail to allow the decision maker to not only assess the strength (i.e., internal validity) of the trial evidence, but also make solid inferences regarding how well the benefits and risks observed in trials will translate into actual practice (i.e., external validity).

In addition to providing a description of individual trial characteristics and results, HTA may also involve the synthesis of trial results for an individual treatment and the indirect comparison of outcomes across treatments. Synthesis can be accomplished through a variety of statistical techniques, often referred to collectively as meta-analysis.[2] It is important to note, however, that meta-analyses typically occur at the level of specific clinical outcomes (e.g., reduction in hemoglobin A1c levels for antidiabetic treatments)[3] rather than the level of integrated B-R measures.

Once the evidence on individual outcomes has been summarized and synthesized, an integration across benefit and risk outcomes must occur, either qualitatively or quantitatively, to inform a proper understanding of a product's overall B-R profile. When done qualitatively, the process may involve a subjective evaluation of summary measures, such as the number needed to treat (NNT) to achieve a beneficial outcome or the number needed to harm (NNH) that results in an adverse outcome.[4] One shortcoming of such an approach is the subjective nature of the task: different evaluators can come to divergent conclusions depending on how they weigh various outcomes (e.g., the relative value of avoiding a stroke versus causing a gastrointestinal bleed when considering an anticoagulant for atrial fibrillation). A second shortcoming is the absence of a common metric to permit direct comparison of the values provided by treatments for different diseases.

The alternative to a qualitative assessment is to integrate benefit and risk quantitatively through the use of a summary metric, such as net health benefit or quality-adjusted life years (QALYs).[5] For example, the benefit associated with a given treatment strategy may be measured in QALYs by assigning to each time period a weight representing the health-related quality of life (QOL) associated with that period, and aggregating those weights over a time span of interest, such as a patient's lifetime. For modeling purposes, the positive value of treatment-associated benefits can be captured in the form of more time spent in more desirable health states (i.e., those with higher QOL weights), while the negative value of treatment-associated risks can be captured through more time spent in less desirable health states (i.e., those with lower QOL weights).

While aggregate measures such as QALYs help address the shortcomings of a qualitative approach, their use requires a formal weighting process based on stakeholder input that introduces its own set of problems. For example, critics have long raised concerns about how well QALYs capture patients' preferences for certain aspects of health beyond those captured in a static description of a health state, such as total time spent in a given health state, or sequence of health states experienced.[6] Critics have also argued that the QOL adjustment used in calculating QALYs may unfairly disadvantage the elderly and those in suboptimal health states compared to younger, healthier individuals. Nevertheless, aggregate measures of health can be indispensable for certain types of economic analyses required by payers in more sophisticated markets, such as the UK, Canada, and Australia.

10.2.3 Geographic Variability in the Application of HTA to Payer Decision Making

In general, countries that rely most heavily on formal HTA processes for access and reimbursement decisions are those with single-payer national insurance systems, such as the National Health Service (NHS) in the UK and Australia's Medicare system. In those societies, there is a tendency to view healthcare as a basic human right, with a corresponding expectation that the government must provide some basic level of healthcare equitably to all citizens. As a consequence, healthcare decision making in these countries is characterized by the formal use of highly structured HTA with explicit and transparent applications of HTA outputs to decision making.

The UK provides a signal example. Manufacturers seeking reimbursement for their products in the UK are required to submit a reimbursement dossier to the UK's National Institute for Health and Care Excellence (NICE) adhering to a publicly available template and including all of the HTA components previously described. The submission and review processes are clearly described through published guidance documents. NICE seeks to answer two questions in any new technology appraisal: How well does the new technology work compared to standard practice in the health service, and how much will adopting the new technology cost compared to standard practice in the health service?

Review of clinical evidence includes formal evaluation of RCT data with consideration of the generalizability of results to the UK population, as well as formal indirect comparison of treatment outcomes with all relevant comparators where head-to-head data are lacking. The review proceeds from a simple premise that healthcare should improve the quality of one's life or increase one's life expectancy. To facilitate this, treatment effectiveness is quantified using the QALY as an aggregate measure of overall benefit.

Through the use of modeling, the expected gain in quality-adjusted life expectancy with the new treatment and the expected added cost (using UK-specific data) are calculated relative to the "next best" treatment option, and these values are combined in a formal (and mandated) cost-effectiveness analysis to create an incremental cost-effectiveness ratio (ICER). The resulting ICER is a key input into NICE's recommendation to the NHS. Since January 2002, it has been mandatory for the NHS in England and Wales to provide funding for medicines and treatments recommended by NICE in its technological appraisal guidance.[7]

In theory, treatments with an ICER below a society's willingness-to-pay threshold are deemed cost-effective and recommended for use, while those above the threshold are rejected. In practice, while there continues to be considerable debate about the most appropriate threshold, treatments with an ICER greater than £20,000–30,000 per QALY are generally not recommended for use.[8] The appeal of this approach is that it allows for a transparent comparison of interventions across all therapeutic areas, thereby permitting a more rational and defensible allocation of a fixed national healthcare budget. While the UK system is idiosyncratic in many ways, it is representative of many countries across the European Union and elsewhere in its explicit use of HTA in coverage determinations.[9]

It should be noted that the reliance on HTA does not end with the initial coverage determination. The evaluation of real-world effectiveness continues after a product is approved for coverage, with the expectation that manufacturers will provide additional real-world evidence that the drug is working as expected based on initial HTA review. Observational studies may be mandated to provide further confidence about long-term safety. In some countries, repricing negotiations are routine and contingent on demonstrated real-world effectiveness and safety. Government-mandated price cuts have become more common as hard economic times have added to downward pressure on national healthcare budgets.

The perspective taken on B-R appraisal and HTA by U.S. payers is different than that taken in non-U.S. HTA systems. In the United States, HTA decision processes are shaped by the view that healthcare is a market-based service subject to supply and demand rather than a universal public entitlement, as it is considered in many economies in Western Europe. In the United States, "economic freedom is an end in itself," and there is strong political pressure encouraging freedom in healthcare choice.[10]

The U.S. healthcare payment system is a mixed one, including both federal and state-funded programs (e.g., Medicare, Medicaid, and TRICARE), private payer insurance systems, and the uninsured (self-pay). With the addition of Part D support for outpatient Medicare prescriptions in 2006 and the passage of the Affordable Care Act in 2010, federal, state and local government spending on healthcare by 2011 had grown to comprise 45% of all US healthcare expenditures.

Despite the need for cost containment in the most expensive healthcare system in the world,[9] the largest payer (i.e., the federal government) has refrained from adopting the kind of analytic tools that integrate cost and effectiveness (such as cost per QALY models) widely used in HTA systems by many other developed healthcare systems. In fact, strong lobbying from many powerful stakeholders has resulted in legislation that places formal limits on the use of cost-effectiveness analysis by the Centers for Medicare and Medicaid Services (CMS), the federal agency that administers the U.S. Medicare program:

> The law allows CMS to use comparative effectiveness research (CER) evidence in coverage and/or reimbursement decisions as long as the coverage process is an iterative one—a standard that the current CMS national coverage determination process meets. CMS may also use CER to establish differential copayments, which could be used in a value-based insurance design program. *The bill specifically prohibits any cost-effectiveness analysis that would use any adjusted life years factor that would place lower value on the life of elderly, disabled, or terminally ill individuals compared to younger and healthier individuals.*[12] (emphasis added)

Thus, comparisons of relative effectiveness and safety (B-R) can be explicitly considered in federal healthcare decisions, but bringing NICE-style cost-effectiveness analyses explicitly into federal payer decisions is prohibited.

Because of the U.S. healthcare system's fragmented nature and resistance to the explicit use of analytic cost-effectiveness tools, the use of HTA by U.S. payers tends to be more variable, and its incorporation into formulary decision making more subjective. The Academy of Managed Care Pharmacy (AMCP) Format for Formulary Submission Dossier version 3.0[13] is the most common format recommended for evidence dossier submissions to payers in the United States. The contents of the dossier generally coincide with the HTA components mentioned previously: product description and place in therapy; available efficacy, effectiveness, and safety information; clinical evidence tables from studies; and economic value evidenced by cost-effectiveness and budget impact models. As elsewhere, RCTs provide the basis for estimates of efficacy and safety for initial formulary reviews, and serve as inputs for systematic reviews, indirect comparisons, meta-analyses, and comparative effectiveness assessments.

The process by which economic assessments are conducted and used in payer decision making in the United States is less prescriptive and formulaic than in the previously discussed single-payer systems. Despite an explicit request for information on cost-effectiveness in the AMCP dossier, there tends to be less emphasis on formal, integrated value assessments and greater emphasis on estimated budget impact. Furthermore, skepticism around manufacturer-produced models has led many payers to develop their own proprietary economic models.

In the United States, *payers* generally refers to teams of professionals within a health plan who jointly determine the relative efficacy, safety, and value of a proposed medication. In some cases, these determinations are made by members of a single pharmacy and therapeutics (P&T) committee, while in others the assessments are divided among multiple teams. For example, a P&T committee might determine whether a product should be added to the formulary based on benefit and risk, while a value assessment committee would negotiate with the manufacturer regarding contract details that affect affordability. Some health plans contract out product analyses and receive recommendations for a preferred drug list (PDL) from companies specializing in formulary benefit design. The health plan P&T committee usually retains the final say on what will be available on the plan's formulary and under what appropriate use restrictions.

This system of access approval in the United States considers a product's perceived relative benefit, risk, and cost. In the absence of clear cost-effectiveness criteria, it is increasingly difficult for payers to refuse access for FDA-approved medications. Rather than defend noncoverage decisions, payers may choose to allow access to patient self-paid treatments (up to 100%) rather than offer them as insured benefits. As opposed to certain non-U.S. HTA systems that limit access to particular medications that do not meet explicit cost-effectiveness thresholds, rationing of expensive healthcare de facto occurs based on ability to pay. In this environment, the weight accorded to the likely benefits and risks of a medication may be dwarfed by the weight accorded to affordability by both health plans and patients.

The post-coverage context of U.S. formulary decision making concerning drug access and restrictions is dynamic and complex. An increasing number of nonindustry, nonregulatory investigators (e.g., payers, academics) are conducting studies using real-world data that utilize administrative claims or electronic medical records data to identify potential safety signals. Some U.S. payers, such as Wellpoint, have instituted mandatory post-action reviews requiring real-world evidence (RWE) of effectiveness in usual care settings.

Despite an aversion for NICE-style cost-effectiveness analysis as a basis for U.S. formulary decisions, cost considerations are pervasive. Sometimes financial concerns overwhelm the consideration of comparative effectiveness and risk. In lieu of accepted criteria for what constitutes effectiveness and safety, clearly articulated cost considerations can be the deciding factor in determining whether and at what stage of treatment patients can have access to medications. Because cost considerations vary from health plan to health plan, the U.S. marketplace provides many choices for healthcare services in line with its preference for economic freedom.

10.3 Recent Trends and Future Opportunities

10.3.1 Trends in Payer Use of Evidence and HTA

In recent years, a number of trends have emerged that are likely to transform the way in which payers conduct and utilize HTA to inform decision making. One such trend has been a movement toward greater harmonization among agencies conducting HTA. A prominent example of this is the formation of the European Union Network for Health Technology Assessment (EUnetHTA), a network of national and regional organizations that produce or contribute to HTAs in Europe. EUnetHTA has adopted, as part of its mission, support for the efficient production and use of HTAs in countries across Europe. This involves not only the sharing of resources, but also the development of common methodological standards and practices.[14] While pilots are already underway to conduct harmonized relative effectiveness assessments, EUnetHTA has recognized that certain elements of HTAs (e.g., cost and economic considerations) cannot be harmonized due to country-specific economic and health system differences. It is worth noting that there has been close collaboration between EUnetHTA and the European Medicines Agency (EMA), including EMA participation in the aforementioned relative effectiveness assessments for pharmaceuticals as well as the provision of joint EMA-HTA scientific advice on clinical trial design to manufacturers. It seems likely that these trends of increased harmonization among conductors of HTA and greater collaboration between regulatory and payer bodies will continue.

A second notable trend has been a growth in payers' interest in conditional reimbursement or "access with evidence development" arrangements. Under these types of agreements, product access and reimbursement is granted conditional upon the generation of additional evidence to address key clinical or economic uncertainties. Such agreements have generally taken one of two forms: (1) "only in research" agreements under which coverage is provided only in the context of a clinical study aimed at reducing a key uncertainty (e.g., long-term safety) or (2) agreements involving an "outcomes guarantee" (either clinical or financial) under which financial accountability is shared between the supplier and purchaser of a treatment, sometimes also referred to as risk-sharing schemes. Stafinski and colleagues, in a comprehensive review of these types of arrangements, documented 32 only-in-research and 26 outcomes guarantee arrangements as of May 2009.[15] The majority of these have been in Europe, and there continues to be greater interest outside of the United States. In the United States, while concepts such as outcomes-based contracting have generated a great deal of discussion, concrete examples are relatively uncommon and exploratory.

While examples of conditional reimbursement arrangements are relatively uncommon in the United States, one established example is the CMS Coverage with Evidence Development (CED) program[16] under which patients may receive services not deemed "reasonable and necessary" on a provisional basis. As a condition for access, research data are collected (e.g., in patient registries) and analyzed to provide insights into the use and experience of patients receiving the test or treatment with goals of gathering more evidence of real-world effectiveness and improving medical practice. As of early 2013, CMS had used the CED program only 19 times, mostly for diagnostics and imaging techniques, and not more than twice for pharmaceuticals.[17]

A third trend of note in the United States has been the rise of RWE in healthcare, sometimes referred to as "big data." Spurred by federal investment in electronic medical records (EMRs), the U.S. healthcare delivery system is building progressively larger databases of high-quality clinical data that are interoperable with related data sets, such as repositories of socioeconomic information and administrative billing records. In addition, there is promise that distributed data systems that are interoperable across healthcare delivery organizations (e.g., Mini-Sentinel)[18] could further add to data capabilities. As is often the case with rapidly developing capabilities like EMRs, there are likely to be more challenges around compatibility of systems than initially anticipated, limiting the ability to easily combine cases across various healthcare systems. Nevertheless, when coupled with "big analytics," these large, integrated data systems ultimately have the potential to provide payers with the kind of high-quality RWE they seek to better evaluate benefit, risk, and cost of new medications in usual care environments—albeit not until after patients have been granted initial access and sufficient evidence has begun to accrue.

10.3.2 Future Opportunities

As noted in Chapters 7 and 8, despite regional differences, coordinated policies coupled with consistent frameworks can improve the harmonization of the B-R approaches, assessments, and communications of regulatory agencies, ultimately improving the quality and dependability of their decision making. In a similar manner, greater harmonization in the manner in which evidence is defined, gathered, and interpreted across payers can improve the quality and consistency of their reimbursement and access decisions. Here, the role of B-R assessment is still evolving. Current industry trends suggest the potential for B-R assessment to play an increasingly prominent role in drug development, and ultimately influence approval and reimbursement decisions. Some degree of harmonization across regulators and payers regarding their demands and expectations of B-R assessments, insofar as such assessments serve as inputs into their own internal deliberations, could provide great value to the patients they both seek to help. Furthermore, clear, consistent, and (where possible) aligned guidance to drug developers could potentially reduce the time and cost of delivering medicines to those in need.

It is likely that improvements in harmonization across payers, regulators, and developers will come initially at the framework level. As highlighted throughout this book (Chapters 1, 5, and 8), a framework is a necessary first step in creating any qualitative or quantitative B-R assessment. Two key elements of a B-R frame are of particular interest to developers, regulators, and payers: perspective and population. Developers typically focus on the degree to which a drug's benefits outweigh its associated risks for a specific patient group under specific conditions and seek supporting evidence by means of clinical trials. Regulators typically focus on the degree to which a drug's benefits and risks extend to a general population. Payers typically focus on the degree to which a drug's benefits and risks will manifest in the specific populations they serve, at an affordable cost. While these frames diverge in perspective and population, the fundamental importance of benefit and risk is consistent across these very different groups.

Enhanced harmonization of B-R frameworks among developers, regulators, and payers will also give an advantage to other stakeholder groups typically removed from the drug development and delivery process. Patient advocates, patient support groups, patient lobbyists, etc., will be better able to undertake and communicate the results of their own B-R assessments of treatments. Increased democratization of technology and continued emergence of easy-to-use tools will further afford these groups the ability to perform their own quantitative assessments. The ease with which B-R assessments can be carried out with ever-greater refinements of perspective and population will eventually converge on the ultimate special interest group: the individual patient. Given the pervasiveness and growth rate of mobile technology, it is quite likely that within a decade tools permitting individualized B-R assessments will be widely available via portable devices during physician office visits or at the point of purchase as a natural part of a drug consultation.[19]

However, there is a fundamental difference between B-R assessment as a means of comparing alternatives and B-R assessment as an input into a larger health technology evaluation at the payer level: the treatment of cost. The B-R frameworks described in this book focus on a single question: Do the benefits promised by a drug adequately compensate the patient for any potential risks incurred as a consequence of its use? While well suited to compare the relative advantages and disadvantages of competing alternative treatments, these frameworks do not address the cost of treatment to the patient or to society, nor do they take into account the patient's or society's willingness to pay for treatment. As emphasized throughout this chapter, these economic issues are of great concern to payers and represent their unique domain. Nevertheless, the commonality of interest in benefit and risk provides an opportunity to meld perspectives among payers and regulators in a way that could create more efficient and consistent decision making among payer communities and between the payer and regulator communities.

10.4 Conclusions

Harmonization among payer communities and between regulators and payers has the potential to increase the efficiency of decision making through the tailoring of information that can fit the purposes of both groups. In addition, payers stand to gain clarity and consistency in their decision making through increased harmonization, the sharing of common perspectives, frameworks for B-R assessments, and rationale for decisions taken. Even so, each community has distinct mandates that limit the degree of harmonization that can and should occur. While payer access and reimbursement decisions need to be informed by clinical benefit and risk, economic considerations should not creep into regulatory decision making. Within payer communities, local preferences for health as a good and the willingness to pay for it suggest that access and reimbursement decisions will remain local even though regulatory approval decisions for a medication may be regional.

Despite the rise of sources of information such as RWE, at the point of market authorization where key access decisions must be made, payers still suffer from inadequate information on how a new medication will fare in usual care practice. To bridge this gap, drug development has begun to turn toward nontraditional (from a regulatory perspective) research designs, such as pragmatic clinical trials, as a way to simulate the usual care environment prior to market authorization. Recently, the head of the FDA's Center for Drug Evaluation and Research suggested that there was a regulatory role for such studies (e.g., to provide long-term safety data).[20]

One thing, however, is clear: while benefit and risk are cornerstones of clinical decision making, affordability is an inescapable constraint. In and of themselves, benefit and risk are insufficient to address the complex healthcare access and cost containment challenges faced by payers; in and of itself, a single-minded focus of affordability will neither aid nor abet the health, welfare, and prosperity of a population. All three elements are crucial: benefit, risk, and cost. And while there is no formula to determine their optimal balance, we can at least be confident that with aligned intent, consistent frameworks and clear communication among all stakeholders, healthcare innovation, technology, access, and policy will continue to improve.

References

1. Committee for Evaluating Medical Technologies in Clinical Use, Division of Health Sciences Policy, Division of Health Promotion and Disease Prevention. 1985. *Assessing medical technologies.* Washington, DC: National Academies Press.
2. Lau, J., Ioannidis, J., and Schmid, C. 1997. Quantitative synthesis in systematic reviews. *Annals of Internal Medicine* 127: 820–826.
3. Sherafali, D., Nerenberg, K., Pullenayegum, E., Cheng, J., and Gerstein H. 2010. The effect of oral antidiabetic agents on A1C levels: A systematic review and meta-analysis. *Diabetes Care* 33: 1859–1864.
4. Sinclair, J., Cook, R., Guyatt, G., Pauker, S., and Cook, D. 2001. When should an effective treatment be used? Derivation of the threshold number needed to treat and the minimum event rate for treatment. *Journal of Clinical Epidemiology* 54: 253–262.
5. Gold, M., Siegel, J., Russell, L., and Weinstein, M., eds. 1996. *Cost-effectiveness in health and medicine.* New York: Oxford University Press.
6. Neumann, P. 2011. What next for QALYs? *Journal of the American Medical Association* 305: 1806–1807.
7. ISPOR Global Health Care Systems road map: United Kingdom (England and Wales)—Reimbursement process. International Society for Pharmacoeconomics and Outcomes Research website. http://www.ispor.org/htaroadmaps/uk.asp.
8. Measuring effectiveness and cost-effectiveness: The QALY. National Institute for Health and Care Excellence website. http://www.nice.org.uk/newsroom/features/measuringeffectivenessandcosteffectivenesstheqaly.jsp.
9. Stabile, M. et al. 2013. Health care cost containment strategies used in four other high-income countries hold lessons for the United States. *Health Affairs* 32: 643–652.
10. Friedman, M. 2002. *Capitalism and freedom.* Chicago: University of Chicago Press.
11. Hartman, M. et al. 2013. National health spending in 2011: overall growth remains low, but some payers and services show signs of acceleration. *Health Affairs* 32: 87–99.
12. PCORI: Dissemination and use of research findings. Center for Medical Technology Policy website. http://www.cmtpnet.org/what-is-cer/pcori/.
13. FMCP Format Executive Committee. 2010. The AMCP format for formulary submissions version 3.0. *Journal of Managed Care Pharmacy* 16 (1 Suppl A): 1–30.
14. EUnetHTA Mission, Vision, and Values. EUnetHTA website. http://www.eunethta.eu/about-us/mission-vision-values.
15. Stafinski, T., McCabe, C., and Menon, D. 2010. Funding the unfundable. *PharmacoEconomics* 28: 113–142.
16. Miller, F., and Pearson, S. 2008. Coverage with evidence development: Ethical issues and policy implications. *Medical Care* 46: 746–751.
17. Neuman, P., and Chambers, J. 2013. Medicare's reset on 'coverage with evidence development.' Health Affairs blog, April 1, 2013. http://healthaffairs.org/blog/2013/04/01/medicares-reset-on-coverage-with-evidence-development/.
18. Mini-Sentinel website. http://www.mini-sentinel.org/.
19. Eckman, M. 2001. Patient-centered decision making: A view of the past and a look toward the future. *Medical Decision Making* 21: 241–247.

20. Woodcock, J. 2013. Comparative effectiveness research and the regulation of drugs, biologics and devices. *Journal of Comparative Effectiveness Research* 2: 95–97.

Epilogue

This book was about benefit and risk.

Now that we have visited one or more of each of the four galleries to which this book offered a gateway, we have gained a sense of the complexities, the numerous important considerations, and the multitude of facets and perspectives that are inherent in a thoughtful analysis of benefits and risks, as a compound moves through its research and development path and into its life cycle as a marketed product. Collectively, the paintings portrayed in the Early Clinical Development, Full Clinical Development, Regulatory Review and Policy, and Post-Launch Assessment galleries paint a picture promising the potential for a richer breadth of interpretation and a deeper understanding of a compound's net therapeutic value than even the most detailed and accurate summary of benefits and risks considered in isolation could offer.

We now find ourselves back in the corridor outside the four galleries and notice a fifth door at its end, presumably leading to one more gallery. However, beyond that door lies an open space that is reminiscent of the previous galleries, but its walls have not yet been finished and there is no ceiling. There is no artwork at all yet in this space, only an easel with a fresh canvas and all the painting utensils needed for continuing the work at hand. How will the future of benefit-risk (B-R) analysis evolve? How will B-R analysis continue to shape the pharmaceutical R&D paradigm, and all the decisions and policies that shape the journey from bench to bedside? There are four considerations in this regard, each with their specific challenges, which have bearing on the responses to these questions.

First, we must acknowledge that significant challenges will persist on the path to widespread adoption of rigorous and systematic B-R analysis. Insofar as widespread adoption will ostensibly imply some kind of agreement on standard practice, or at least adherence to a common set of principles by all the various stakeholders involved in developing a drug and bringing it to patients, the very nature of B-R analysis suggests that this will be difficult to achieve. We have seen throughout this book that the thoughtful assessment of B-R is not a "one size fits all" proposition. Varied perspectives must be taken into account. Numerous approaches can be leveraged to gain useful insights. Some stakeholders, most notably payers, must broaden the B-R scope to take economic considerations into account, whereas for others—regulators, for instance—these are justifiably irrelevant. There is no single answer. In view of the multifaceted complexity that is inherent and defining for B-R analysis, attempts to standardize could in fact easily have a negative and limiting impact. Articulation of what constitutes best practice is understandably difficult within any single stakeholder group (pharmaceutical companies, regulatory bodies, payers, prescribers, patients), and harder still across stakeholders. The magnitude of this challenge may be appreciated by considering the significant effort typically required by a single regulatory agency to produce a guidance document on what is often an important but relatively narrowly focused aspect of drug development. In contrast, the challenge of broad harmonization in the practice of B-R assessment seems enormous. It appears that a crucial step toward this goal will involve the formulation and adoption of foundational principles that will serve as formal guideposts for developing, expanding, and leveraging B-R frameworks in drug development.

Second, the full value of B-R analysis will not be realized without a significant shift toward appropriate, fit-for-purpose quantitative assessment. This, too, presents a difficult challenge, as Chapter 8 has suggested. In many cases, the customer of B-R evaluations will have to become comfortable with shifting his or her judgment from a largely subjective basis to a more analytical basis. This will require the development of a fundamental trust in the underlying analytics before one can even begin to interpret a quantitative B-R assessment and have the kinds of conversations that it is intended to foster. The skepticism toward quantitative B-R evaluations is rooted in not only the fact that many are unaccustomed to interpreting such evaluations, but also often a basic mistrust in the methodologies used to obtain them— certainly this skepticism speaks to a lack of understanding of the value of the quantitative methods in question. Nonetheless, the kind of evolution that, for example, statistical methodology for clinical trial design has undergone over the past six decades offers encouragement that similar progress in the integrated evaluation of benefits and risks is also possible. For instance, it is possible for a statistician and a clinician to debate the appropriate assumptions that affect the sample size calculation of a clinical trial without the latter understanding the complex mathematical details required for the calculation itself. The point is that the clinician trusts that the sample size is determined according to a rigorous methodology that need not be debated, clearing the path for what should rather be the topic of debate, namely, choice of effect size, variance assumptions, and other parameters that may affect sample size. Getting to a similar point in B-R analysis will require increased use and new applications of statistical methods with sound properties that are well studied and understood. Bayesian statistics, in particular, hold a great deal of potential for quantifying a changing B-R balance over time, in terms of the ability to assimilate disparate sources of prior information (including qualitative forces such as biases and motivations) and update assessments based on new data. Up to the present time, the limited statistical representation in the conversations advancing the frontiers of B-R analysis in drug development has been noticeable (as witnessed, for instance, by the lack of statistician contributors to this book).

Solidifying the partnership between statisticians and their colleagues from other disciplines in evolving the analytical aspects of B-R assessment will be indispensable. This leads to the third consideration around the broader question of how the future of B-R analysis will evolve: the need for education. Whereas the statistics community is not the only professional function within the pharmaceutical industry that has a significant ownership stake in advancing methods for quantitative B-R assessment, its role in this regard is indispensable and expected to increase. To facilitate this development, statisticians will have to broaden their collaborative networks and understand the opportunities as well as the practical limitations for increased analytics in the short term, with a view toward helping define a realistic vision for the future, in terms of broadening the use and impact of more sophisticated B-R analyses. As far as the ground-level implementation of B-R assessments within companies is concerned, proponents within statistics, decision sciences, drug safety, or even dedicated B-R organizations will need to add numbers to their ranks—especially drug development experts with keen facilitation skills, who are at ease communicating across disciplines, who are not stifled by uncertainty, who can foster dialogue, and who can weave a highly heterogeneous collection of inputs into a coherent story.

Finally, as the conversation around the current role and future potential of B-R analysis continues to evolve, one must keep in mind always that it is a conversation between many artists who are jointly working on a painting. It represents a collaborative effort between scientists, statisticians, clinicians, regulators, policy makers, and economists. It involves patients, prescribers, and payers. This collaboration implies that progress toward the ideal future state of B-R analysis will necessarily be slowed by the practical constraints that broad groups face when working together toward a common goal. Yet this collaboration is indispensable if the unified assessment of benefits and risks is to become a durable, useful norm, and the lens through which the fruits of pharmaceutical R&D are viewed. Happily, the journey is well underway.

Thus, we have discovered that it no longer suffices to paint two pictures, one of benefits, the other of risks, and to ask the viewer to bring the two together. There is only one canvas.

This book was about *benefit and risk*.

Andreas Sashegyi
James Felli
Rebecca Noel
Indianapolis, April 2013

Glossary

AE: Adverse event
AMCP: Academy of Managed Care Pharmacy
ANDA: Abbreviated new drug application
BLA: Biologic license application
B-R: Benefit-risk
BRAT: Benefit-Risk Action Team
CBER: Center for Biologics Evaluation and Research
CDER: Center for Drug Evaluation and Research
CDS: Core data sheet
CED: Coverage with evidence development
CHMP: Committee for Medicinal Products for Human Use
CI: Confidence interval
CIOMS: Council for International Organizations of Medical Sciences
CIRS: Center for Innovation in Regulatory Science
CMR: Centre for Medicines Research
CMS: Centers for Medicare and Medicaid Services
COBRA: Consortium on Benefit-Risk Assessment
cRMP: Core risk management plan
CV: Cardiovascular
DDI: Drug-drug interaction
DIA: Drug Information Association
EFPIA: European Federation of Pharmaceutical Industries and Associations
EMA: European Medicines Agency
EMR: Electronic medical records
EPAR: European Public Assessment Reports
EUnetHTA: European Union Network for Health Technology Assessment
FDA: U.S. Food and Drug Administration
FDAAA: Food and Drug Administration Amendments Act
FDAMA: Food and Drug Administration Modernization Act
FDASIA: Food and Drug Administration Safety and Innovation Act
FE: Favorable effect
HHS: Health and Human Services
HTA: Health technology assessment
ICER: Incremental cost-effectiveness ratio
ICH: International Conference on Harmonization
IMI: Innovative Medicine Initiative
INR: International Normalized Ratio
KBRS: Key benefit-risk summary
LFT: Liver function test

MAB: Minimal acceptable benefit
MAR: Maximal acceptable risk
MCDA: Multicriteria decision analysis
MI: Myocardial infarction
MTX: Methotrexate
NCA: National Competent Authority
NDA: New drug application
NHS: National Health Service
NICE: National Institute for Health and Care Excellence
NNH: Number needed to harm
NNT: Number needed to treat
P&T: Pharmacy and therapeutics
PCI: Percutaneous coronary intervention
PDUFA: Prescription Drug User Fee Act
PhRMA: Pharmaceutical Research and Manufacturers of America
PrOACT-URL: Problem Formulation, Objectives, Alternatives, Consequences, Trade-Offs, Uncertainties, Risk Attitude, and Linked Decisions
PROTECT: Pharmacoepidemiological Research on Outcomes of Therapeutics by a European Consortium
PSUR: Periodic safety update report
QALY: Quality-adjusted life year
QOL: Quality of life
Q-TWiST: Quality-adjusted time without symptoms or toxicity
RCT: Randomized controlled trial
R&D: Research and development
REMS: Risk Evaluation and Mitigation Strategy
RWE: Real-world evidence
SAE: Serious adverse event
SMAA: Stochastic multicriteria acceptability analysis
SOC: Standard of care
TPP: Target product profile
TWiST: Time without symptoms or toxicity
UFE: Unfavorable effect

Index